Artificial Intelligence, Machine Learning and Blockchain in Quantum Satellite, Drone and Network

Quantum Computing is a field in which advanced technologies like quantum communication, artificial intelligence and machine learning can be used to secure and speed up connectivity using quantum computers, quantum drones or quantum satellites. This book serves as a foundation for researchers and scientists in this field. Future technologies, such as quantum drone delivery systems, quicker internet and climate change mitigation, will need quantum information processing and quantum computation. This book deeply explores the importance of quantum computing in real-time applications. It may be used as a reference book for students in higher education, including undergraduate and graduate students, as well as researchers.

Key Features:

- Provides a clear insight into the Internet of Drones for academicians, postdoc fellows, research scholars, graduate and postgraduate students, industry fellows and software engineers
- Useful to professionals who seek information about the Internet of Drones including experts in quantum computing and physics and post-quantum cryptography, as well as data scientists and data analysts
- Covers quantum computing and security for Unmanned Aerial Vehicles (UAV) or drones, which are widely useful for applications related to military, government and non-government systems
- Explores futuristic aspects of the Intenet of Drones to improve everyday living for ordinary people

Artificial Intelligence, Machine Learning and Blockchain in Quantum Satellite, Drone and Network

Edited by
Thiruselvan Subramanian, Archana Dhyani,
Adarsh Kumar and Sukhpal Singh Gill

CRC Press
Taylor & Francis Group
Boca Raton London New York

CRC Press is an imprint of the
Taylor & Francis Group, an **informa** business

First edition published 2023
by CRC Press
6000 Broken Sound Parkway NW, Suite 300, Boca Raton, FL 33487-2742

and by CRC Press
4 Park Square, Milton Park, Abingdon, Oxon, OX14 4RN

CRC Press is an imprint of Taylor & Francis Group, LLC

Library of Congress Cataloging-in-Publication Data
Names: Subramanian, Thiruselvan, editor.
Title: Artificial intelligence, machine learning and blockchain in quantum satellite, drone and network / edited by Thiruselvan Subramanian, Archana Dhyani, Adarsh Kumar and Sukhpal Singh Gill.
Description: First edition. | Boca Raton, FL : CRC Press, 2023. |
Includes bibliographical references and index.
Identifiers: LCCN 2022015752 (print) | LCCN 2022015753 (ebook) |
ISBN 9781032168036 (hbk) | ISBN 9781032168050 (pbk) |
ISBN 9781003250357 (ebk)
Subjects: LCSH: Quantum computing. | Artificial intelligence. |
Blockchains (Databases) | Machine learning. | High altitude platform systems (Telecommunication) | Artificial satellites in telecommunication.
Classification: LCC QA76.889 .A78 2023 (print) | LCC QA76.889 (ebook) |
DDC 006.3/843—dc23/eng/20220623
LC record available at https://lccn.loc.gov/2022015752
LC ebook record available at https://lccn.loc.gov/2022015753

ISBN: 9781032168036 (hbk)
ISBN: 9781032168050 (pbk)
ISBN: 9781003250357 (ebk)

DOI: 10.1201/9781003250357

Typeset in Times
by codeMantra

Contents

Editors

Thiruselvan Subramanian, PhD, is currently working as assistant professor at Presidency University, Bengaluru. He obtained a graduate degree in Computer Science from Bharathidasan University Tiruchirappalli, India. He received a postgraduate degree and PhD in Cloud Computing from the Department of Computer Applications, National Institute of Technology, Tiruchirappalli, India. He worked at Softeon, Inc., Chennai, as project trainee from February 2010 to June 2010 and then as a software engineer from June 2010 to August 2011. He also worked in Vignan's Foundation for Science, Technology & Research, Guntur as an assistant professor from March 2017 to April 2019. His research interests include cloud computing, decision theory and parallel programming.

Archana Dhyani, PhD, is an assistant professor at the University of Petroleum and Energy Studies, Dehradun, India. She received her PhD in physics from G.B. Pant University of Agriculture and Technology, Pantnagar, India in 2009. She has authored several publications including articles in reputed journals and chapters in standard books, etc. She has also been an external expert member of the National Science Center, Poland and a lifetime member of the Indian Physics Association (IPA). She is also on the reviewer panel of many reputed journals. Her current research interest is to study the spectral and transport properties of quantum dot-based nano-junctions for their various applications like quantum computation, etc.

Adarsh Kumar, PhD, is an associate professor at the School of Computer Science with the University of Petroleum & Energy Studies, Dehradun, India. He received his Master's degree (M. Tech.) in Software Engineering from Thapar University, Patiala, Punjab, India and earned his PhD degree from Jaypee Institute of Information Technology University, Noida, India followed by Postdoc from SRI, AIT, Ireland. From 2005 to 2016, he had been associated with the Department of CSE, JIIT, India, where he worked as assistant professor. His main research interests are cybersecurity, cryptography, network security, and ad hoc networks. He has written many research papers in reputed journals, conferences and workshops. He participated in one European Union H2020-sponsored research project, and he is currently executing two research projects sponsored by the UPES SEED division and Lancaster University, respectively.

Sukhpal Singh Gill, PhD, is a lecturer (assistant professor) in Cloud Computing at the School of Electronic Engineering and Computer Science, Queen Mary University of London, UK. Prior to his present stint, Dr. Gill has held positions as a research associate at the School of Computing and Communications, Lancaster University, UK and also as a post-doctoral research fellow at CLOUDS Laboratory, The University of Melbourne, Australia. Dr. Gill is serving as an associate editor in Wiley ETT and IET Networks Journal. He has co-authored 70+ peer-reviewed papers (with H-index 30+) and has published in prominent international journals and conferences such as IEEE TCC, IEEE TSC, IEEE TII, IEEE IoT Journal, Elsevier JSS and IEEE CCGRID. He has received several awards, including the Distinguished Reviewer Award from SPE (Wiley), 2018, Best Paper Award AusPDC at ACSW 2021 and has also served as the PC member for venues such as PerCom, UCC, CCGRID, CLOUDS, ICFEC, AusPDC. Dr. Gill served as a guest editor for SPE (Wiley) and JCC Springer Journal. He is a regular reviewer for IEEE TPDS, IEEE TSC, IEEE TNSE, IEEE TSC, ACM CSUR and Wiley SPE. He has edited a research books for Elsevier, Springer and Taylor & Francis Group. Dr. Gill has reviewed 400+ research articles of high ranked journals and prestigious conferences according to the data from Publons. His research interests include cloud computing, fog computing, software engineering, Internet of Things and energy efficiency. For further information, please visit http://www.ssgill. me.

Contributors

Rupam Bhagawati
Department of Computer Science and
  Engineering
Presidency University
Bengaluru, India

Chirag Dhara
School of Interwoven Arts and Sciences
Krea University
Sri City, India

Santosh R. Gaikwad
Department of MBA
MET's Institute of Management
Nashik, India

Sukhpal Singh Gill
School of Electronic Engineering and
  Computer Science
Queen Mary University of London
London, United Kingdom

Harpreet Kaur
Department of Computer Science and
  Engineering
Punjabi University
Patiala, India

Harveen Kaur
PG Department of Computer Science
G.S.S.D.G.S. Khalsa College
Patiala, India

Navjot Kaur
Department of Computer Science and
  Engineering
Punjabi University
Patiala, India

Prabhsharan Kaur
Department of Computer
  Science
Chandigarh University
Mohali, India

Adarsh Kumar
Department of Systemics, School of
  Computer Science
University of Petroleum and Energy
  Studies
Dehradun, India

Ajay Kumar
Department of Civil Engineering
G.B. Pant Institute of Engineering and
  Technology
Pauri Garhwal, India

Mahalingam P.R.
Pre Sales Consultant
InApp Information Technologies Pvt.
  Ltd.
Thiruvananthapuram, India

Manasa C.M.
Department of Computer Science and
  Engineering
Presidency University
Bangalore, India

P. Mandal
Department of Physics,
  Cluster of Applied Sciences,
  School of Engineering
University of Petroleum and Energy
  Studies
Dehradun, India

Naga Raju Mysore
Department of Computer Science and
    Engineering
Presidency University
Bengaluru, India

Kamal Pandey
Indian Institute of Remote
    Sensing
Dehradun, India

Pavithra N.
Department of Computer Science and
    Engineering
Presidency University
Bengaluru, India

Arish Pitchai
Data Scientist, Research and
    Development
Entropik Technologies Pvt. Ltd
Chennai, India

S. Poornima
Department of CSE – SOE
Presidency University
Bengaluru, India

Preethi
Department of Computer Science and
    Engineering
Presidency University
Bengaluru, India

Prashant Rawat
Department of Physics,
    Cluster of Applied Sciences,
    School of Engineering
University of Petroleum and Energy
    Studies
Dehradun, India

Sapna Renukaradhya
Department of Computer Science and
    Engineering
Presidency University
Bengaluru, India

Isha Sharma
Department of Computer Science
Chandigarh University
Mohali, India

Surbhi Sharma
School of Computer and Systems
    Sciences (SCSS)
Jawaharlal Nehru University
New Delhi, India

Manmeet Singh
Jackson School of Geosciences
The University of Texas at Austin
Austin, Texas
Ministry of Earth Sciences
Indian Institute of Tropical Meteorology
Pune, India

Rahul Kumar Singh
School of Computer Science
University of Petroleum and Energy
    Studies
Dehradun, India

Thiruselvan Subramanian
Department of Computer Science and
    Engineering
Presidency University
Bengaluru, India

Bijjahalli Sadanandamurthy Sushma
Department of IT, School of CS & IT
Jain (Deemed to be University)
Bengaluru, India

B.S. Tewari
Department of Applied Sciences and
    Humanities
G.B. Pant Institute of Engineering and
    Technology
Pauri Garhwal, India

Steve Uhlig
School of Electronic Engineering and
    Computer Science
Queen Mary University of London
London, United Kingdom

Taskeen Zaidi
Department of IT, School of CS & IT
Jain (Deemed to be University)
Bengaluru, India

1 Quantum Information Processes
Role of Quantum Logic Gates

B.S. Tewari
G.B. Pant Institute of Engineering and Technology

P. Mandal and Prashant Rawat
University of Petroleum and Energy Studies

CONTENTS

1.1 INTRODUCTION

A quantum computer uses quantum circuits, which are made up of wires and quantum gates [1]. Quantum logic gates are used to couple qubits for getting entangled states and long coherent states required for quantum information manipulation. The quantum information is moved around in the wires and then manipulated using quantum gates. The quantum information is then turned into classical information at the end for its utilization.

The idea of quantum computation was first introduced in 1980, well before the realization of the concept of a quantum computer. It was a theoretical concept at that time and focused on the possibility of the realization of the ultimate limits of computation. The situation was like that until 1994, when Shor made a breakthrough and quantum computing became a reality due to his proposed algorithm for

DOI: 10.1201/9781003250357-1

factorizing large numbers that were beyond the efficiency of any classical algorithm [2]. This development laid the foundation for the feasibility of quantum computing in the coming years. Following this development of the possibility of quantum computation, Deutsch [3] published a landmark paper in 1985. That paper highlighted the major issues that lay on the way to bringing into existence a quantum analogy to the Turing machine. These issues were regarding the storage (or memory) and processing that would be required of such a possible physical machine like a quantum computer.

The present chapter explores the working and role of quantum logic gates in quantum information manipulation. The chapter is built in terms of the following subparts: first, an introduction to quantum computation necessary to understand the subject matter; second, an introduction to qubits and quantum entanglement; third, an understanding of logical operations due to quantum gates to manage quantum entanglement for quantum computing; fourth, quantum information processes and the related algorithms; and, fifth, quantum computation relevant to satellite communication.

1.2 LITERATURE SURVEY

Researchers around the world have been working since the end of the last century on the development and applications of quantum computation. Shor was the first to introduce, in 1994, the quantum algorithm required for quantum computation [2,4]. His idea of quantum algorithm was based on efficient factorization of large numbers to encrypt information/message sent over a channel using quantum Controlled-NOT gate. The next advancement was demonstrated of quantum algorithm [5]. These authors developed the probabilistic logic gates and their algorithm based on the concept of polarization beam splitter using probabilistic manipulation of entangled photons developed by Knill et al. [6] and Koashi et al. [7] in the same year.

The first experimental photon-based quantum-controlled NOT gate was demonstrated in the year 2004 [8]. Different sets of entangled ancillary pairs of photons and their inherent interactions were used to develop that gate. This step towards the realization of a quantum-based decision-making machine has opened up new paths for future developments. Apart from the photon-based quantum logic and algorithms, there has been another concept that has been found useful in the development of logic operations. Another idea had also been reported on quantum computation in 2005 [9] employing a cluster state based on Raussendorf's notions [10]. Nielsen's approach is applicable to any non-trivial linear optical gate that passes with a certain probability but has an effect on a calculated measurement when it fails. A logic gate based on optical entanglement was also proposed. It was designed with the help of partially polarizing beam splitters [11,12]. Quantum Process Tomography is also used to demonstrate that a controlled 'NOT' gate can be operated with both a continuous signal and a pulsed signal [12]. Langford et al. also developed a bell state analyser that is non-deterministic and completely resolving by utilizing this gate. This design has been shown to be quite promising for those working in the field of quantum optics. In 2006, appeared a detailed view of quantum information processing [13]. Next, Kok et al. [14] proposed a mathematical approach for studying entanglement in arbitrary coordinate transformations based on quantum field theory. Those authors also reviewed the work done on linear quantum computing in the succeeding year [15].

In the succeeding years, up to 2013, research had been focused around the development of quantum logics, which were based on the most recognized approach, the use of polarized entangled photons as qubits. Cai et al. [16] developed a quantum logic gate-based approach for solving a two-variable problem [16,17]. They used four quantum and four logic gates for the implementation of subroutines. Researchers [18], in 2015, presented a reprogrammable photonic quantum circuit that can take up to six photons as input and has a photon detector circuit for logic operations. That development is another pathway to studying a variety of quantum computing applications. A few reviews have been published that focus on various methods developed in these years [19–21].

In recent years, a lot of research has been observed towards the development of quantum logic circuits and, hence, quantum computation. Wang et al. [22] reported the interfacing of quantum systems via classical channels, demonstrating a satisfactory improvement over prior models. In addition, Schafer et al. [23] demonstrated fast and resilient two-qubit gates for trapped-ion qubits, and all of those techniques revealed highly promising findings for the construction of optical integrated circuits. With each passing year, quantum computing is revealing new perspectives of applications such as data analysis [24] through high-speed processors, overcoming the constraint of information security through blind quantum computing [25]. Studies have also led to the development of a nine-qubit device for error correction applications [26]. Quite recently, a survey on different theoretical and experimental approaches of quantum networks and algorithms and their application for quantum satellite and drone technologies has been performed, and it presents the importance of quantum computation in this direction, which opens doors for enormous applications [27]. The next sections in this chapter are devoted to explaining important quantum gates and their operations in satellite communications.

1.3   QUANTUM BITS AND QUANTUM ENTANGLEMENT

Unlike, and in contrast to the classical computers, the memory units for quantum computers have been one of the main concerns for scientists and engineers. Yet another point of concern has been the process of computation itself, due to the involvement of many quantum-based phenomena. A two-level quantum system can be considered for quantum computing, in analogy to the classical computation based on bits. This system produces a two-valued quantum variable, commonly known as qubits (quantum bits) with its eigenvalues labelled as $|0>$ and $|1>$. These $|0>$ and $|1>$ resemble the classical bits '0' and '1'. The difference between a classical bit and a qubit is that a classical bit possesses a pure discrete value, while a quantum bit or qubit is a quantum state and can also be represented as a superposition of two states. A qubit can generally be represented by the state $\{a_0|0> + a_1|1>\}$, where a_0 and a_1 are coefficients, which are complex in nature and normalized to 1. Thus, besides the two discrete eigenstates, a qubit can be in any possible state of a continuous range of superposition states based on those two discrete states. Hence, an infinite number of states are possible to a qubit,

TABLE 1.1

Differences between a Quantum Bit (Qubit) and a Classical Bit

Qubit	Classical Bit
It reduces to classical bit upon measurement.	It remains unique and unchanged even upon measurement.
A superposition of two-qubit states is also a qubit state.	No superposition state exists.
It possesses entanglement or correlation behaviour.	No entanglement or correlation behaviour exists.
Teleportation is possible with the help of qubit.	Teleportation cannot be possible.

out of which only two states are linearly independent or orthogonal in nature. This is the primary difference between classical and quantum memory units.

The next step in quantum computing is the process of reading information from a qubit. This aspect is known as measurement collapse in quantum mechanics. As the name suggests, 'measurement collapse' represents the loss of quantum character of a qubit as it reduces it to a classical bit during measurement or reading of a quantum bit (qubit). This indicates that upon measurement of a qubit, only one or two eigenstates |0> and |1> are read out of an infinite number of available superposition states.

In order to understand the computation in an n qubit quantum computation, consider initially a basic quantum system of two qubits. These states can be expressed as product eigenstates as per their property. Classical logic for a two-bit system possesses |00>, |01>, |10> and |11> basis states. In contrast to the classical logic system, these qubits can interfere with one another to form $a_{00}|00>+a_{01}|01>+a_{10}|10>+a_{11}|11>$ type of additional superposition states and results in $2n$ product of available eigenstates in their Hilbert space. Table 1.1 presents a comparison of the classical bit and qubit systems.

1.4 QUANTUM COMPUTATION: QUANTUM LOGIC GATES

The working of a quantum computer is based on quantum information processes and manipulation [28]. After the understanding of memory units of a quantum computer, the next significant point of understanding is the computation methodology of quantum computers, i.e., the manner and type of operations that can be performed to solve problems using the concept of quantum mechanics. The quantum logic gates manipulate quantum information and perform operations that transform qubit states. Hence, the understanding of logic operations is an important task to understand the working of a quantum computer. Quantum logic gates are classified into two categories based on the number of qubit operations. Single qubit quantum gates operate with a single qubit, whereas multiple qubit quantum gates work with more than one qubits. One of the most essential characteristics of quantum gates is that they must be invertible, which means that we can know the input once we get the output. Matrixes are used to represent quantum gates. The matrix U that describes the gates must be unitary, that is, it must meet the following conditions:

$$U^+U = I \tag{1.1}$$

The adjoint of U is U^+, and I is a unitary matrix. The necessity to conserve probability imposes a limitation on being unitary. The sole limitation on quantum gates is that they must be unitary; otherwise, any unitary matrix can be a quantum gate. One of the simplest ways to generate a unitary operator in quantum mechanics is to consider the time evolution of a quantum system and put the Hamiltonian in an exponent.

$$U = e^{[i\int_{t_i}^{t_f} H(t)dt]} \tag{1.2}$$

Hence, the processing of quantum computation may be modelled in the framework of unitary operations between input and output qubits. To handle this concept, many possibilities towards the physical realizations have been proposed in the form of cavity-Quantum Electrodynamics, trapped-ions, solid-state and superconducting-quantum dot junctions. A block diagram for such an operation can be drawn as in Figure 1.1.

From the above discussion, it is clear that unitary operations are in the central role in quantum computation operations. Therefore, quantum computation operation possesses a reversible process, in a logical and physical way, due to the inherent property of unitary operation. Here, one can also add that all reversible Boolean functions and their operations can be placed as a special case of unitary transformations. In other words, any problem that can be classically simulated may also be quantum mechanically simulated. Further, the additional properties of quantum computers, viz. superposition, entanglement, etc., make them exponentially quicker than any conventional computer. Few other tasks such as quantum teleportation can also be achieved with the help of quantum computations. In practice, it is a well-known fact that the input and output for any quantum computer will be classical in nature. Hence, there is a basic requirement of any quantum computer to have a quantum algorithm which can read classical inputs and provide classical output again for efficient read-out after classical or quantum processing or computation. Straightforward quantum algorithms and operations for single-qubit and multi-qubit quantum logic gates are discussed below.

1.4.1 SINGLE-QUBIT GATES

The operation of a single-qubit quantum gate can be understood with the example of a simple NOT gate. A classical NOT gate interchanges the 0 and 1 states. In the case of quantum NOT gate, a linear combination of states $|0>$ and $|1>$ transforms linearly as $a_1|0>+a_2|1>$ to $a_1|1>+a_2|0>$ [29]. In terms of matrices, these transformations of quantum NOT gate can be represented as below:

$$X \begin{bmatrix} a_1 \\ a_2 \end{bmatrix} = \begin{bmatrix} a_2 \\ a_1 \end{bmatrix}$$

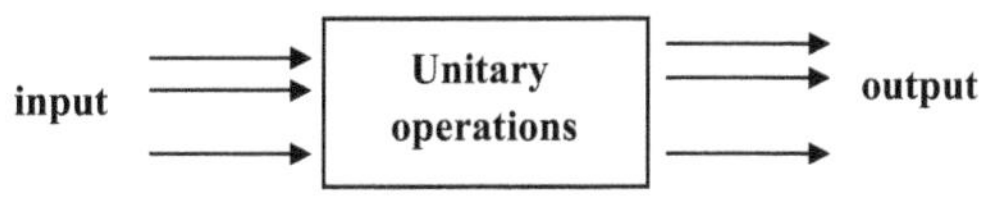

FIGURE 1.1 Block diagram of quantum computation based on unitary operations

where X is a Pauli matrix and can be expressed as below:

$$X = \begin{bmatrix} 0 & 1 \\ 1 & 0 \end{bmatrix}$$

Other forms of Pauli matrices or quantum gates based on different terms can also be expressed as below:

$$Y = \begin{bmatrix} 0 & -i \\ i & 0 \end{bmatrix}$$

$$Z = \begin{bmatrix} 1 & 0 \\ 0 & -1 \end{bmatrix}$$

One of the most useful Pauli matrices for handling quantum computation is the Walsh-Hadamard gate, which can be expressed as below:

$$WH = \frac{1}{\sqrt{2}} \begin{bmatrix} 1 & 1 \\ 1 & -1 \end{bmatrix}$$

A Walsh-Hadamard (WH) quantum gate has no classical analogue. Its operation is to transform $|0>$ into $\frac{1}{\sqrt{2}}(|0\rangle + |1>)$ and $|1>$ into $\frac{1}{\sqrt{2}}(|0\rangle - |1>)$. That quantum gate thus converts each eigenstate into a combination of more than one eigenstates with alternating phases [29,30]. The last single-qubit gate is the phase shift gate [31]. It is represented by $\emptyset$. The matrix representation of the phase shift gate is as follows:

$$\emptyset = \begin{bmatrix} 1 & 0 \\ 0 & e^{i\emptyset} \end{bmatrix}$$

1.4.2 MULTI-QUBIT QUANTUM GATES

To achieve multi-qubit gates, entanglement amongst the qubits is essential. The single-qubit quantum gates mentioned in the above section are not sufficient for performing all types of unitary operations. This indicates a necessity for universal quantum gates that can generate all unitary matrices. Many works have been reported in this direction. It is not surprising that three-qubit quantum gates were proposed by Deutsch [32], prior to the development of a two-qubit gate. This is because three-bit gates are requisite for universal reversible operations in classical computation. Deutsch identified a three-qubit quantum gate by generalizing a three-bit classical gate known as Toffoli gate. Toffoli gate is a universal gate which can perform reversible Boolean logics. A representation of the three-bit Deutsch gate is given below.

$$D(\alpha) = \begin{bmatrix} 1 & & & & & & & \\ & 1 & & & & & & \\ & & 1 & & & & & \\ & & & 1 & & & & \\ & & & & 1 & & & \\ & & & & & 1 & & \\ & & & & & & i\cos\alpha & \sin\alpha \\ & & & & & & \sin\alpha & i\cos\alpha \end{bmatrix} \begin{matrix} |000\rangle \\ |001\rangle \\ |010\rangle \\ |011\rangle \\ |100\rangle \\ |101\rangle \\ |110\rangle \\ |111\rangle \end{matrix}$$

where parameter α is an irrational multiple of π. The Deutsch gate consists of the following property:

$$D(\alpha)D(\alpha') = iD(\alpha + \alpha') \tag{1.3}$$

The Deutsch gate is able to perform any unitary operation for any number of input qubits by repeated applications of these qubits at a time.

A few of the issues, for example, like that of universality in classical computation may exist for quantum computation too. After almost 6 years of the introduction of three-qubit gates, two-qubit gates were developed [33–35]. It has been observed that a three-bit Toffoli gate can be achieved with the help of two-qubit gates following quantum concepts. There are several ways to build the two-qubit gates (special cases are single-qubit unitary relations). Similar gates can be built using three parameters [35]. To demonstrate this, consider that with the exception of an overall phase component, a general way of expressing a two-dimensional unitary matrix is as follows [36]:

$$y(\lambda, v, \phi) = \begin{bmatrix} \cos\lambda & -e^{iv}\cos\lambda \\ e^{i(\phi-v)}\sin\lambda & e^{i\phi}\cos\lambda \end{bmatrix} \tag{1.4}$$

All single-qubit transformations, including the gates x and z, are special instances of Eq. (1.4). To demonstrate how this works, consider the fact that y has the attribute of being able to translate any arbitrary qubit state to the eigenstate $|1\rangle$. Choosing $\cos\lambda = |c_0|$, $v = \arg\dfrac{c_0}{c_1}$ and $\phi = \arg\{c_1\}$ for a state, $c_0|0\rangle + c_1|1\rangle$ provides the transformation, $y(\lambda, v, \phi) : c_0|0\rangle + c_1|1\rangle \rightarrow |1\rangle$. Since y is unitary and reversible, all states of the qubit can be obtained from any other of its state using y. In other words, y is sufficient to generate all single-qubit unitary operations. y may be used to construct a family of universal two-qubit gates.

$$\Delta_2[y] = \begin{pmatrix} 1 & 0 & 0 & 0 \\ 0 & 1 & 0 & 0 \\ 0 & 0 & \cos\lambda & -e^{iv}\cos\lambda \\ 0 & 0 & e^{i(\phi-v)}\sin\lambda & e^{i\phi}\cos\lambda \end{pmatrix} \tag{1.5}$$

The notation $\Delta_2[y]$ indicates that this is a two-qubit gate that applies y to the second qubit if the first qubit is in $|1>$. It is worth noting that $\Delta_2[y]$ includes y, since an operation on a single qubit can be done conditionally on another qubit first being in $|0>$ and subsequently in $|1>$. Equation (1.5) also includes a graphical representation of $\Delta_2[y]$. The set of gates, $\Delta_2[\lambda, \upsilon, \phi]$, is universal for quantum logic in the sense that it may mimic any unitary transform on any number of qubits by operating on just two qubits at a time. It can be seen easily that the operations of the first two states are on $|00>$ and $|01>$. This indicates that the above matrix performs as a two-dimensional matrix instead of a four-dimensional matrix.

1.5 QUANTUM INFORMATION PROCESSES AND ALGORITHMS

The quantum information processes are held in quantum circuits, which can be built from quantum gates. The quantum circuits are obtained by arranging the quantum gates in a particular fashion so that it is possible to execute a particular type of quantum algorithm, wherein running the quantum gates is actually a unitary transformation that acts on a single qubit or pair of qubits. A measurement determines a traditional result such as 0 or 1 at the end of the process. The typical operations of quantum logic gates, which are discussed in the previous section and used in quantum circuits, are based on the following postulates of quantum informatics [37]:

> **First Postulate:** The first postulate defines the state space with a state vector (unit length and complex coefficients) in a Hilbert space.
> **Second Postulate:** In the second postulate, the evolution of a closed system is defined by means of unitary transforms.
> **Third Postulate:** The third postulate is responsible to relate the measurements of quantum system and classical world.
> **Fourth Postulate:** Composite systems are specified in the last postulate.

There are many interesting quantum algorithms, which can handle a range of practical problems. These problems are solvable more efficiently using quantum algorithms, in contrast to classical computing algorithms. The quantum algorithms are broadly divided into three categories. The first is a group of algorithms based on quantum analogues of the Fourier transform. The examples are Deutsch's algorithm and Shor's algorithm. The second and third classes of algorithms are quantum search algorithms and quantum simulation algorithms, respectively, in which a quantum computer is used to search data and simulate a quantum system, respectively. The first class of algorithms have been reported accompanying a giant development in quantum circuits and computations.

To help understand the working of quantum algorithms, the Deutsch algorithm is discussed here. This algorithm shows how quantum computation can solve classical computational problems more efficiently than classical computers [38]. As we know from the previous section, Walsh-Hadamard (*WH*) quantum gate is a very important

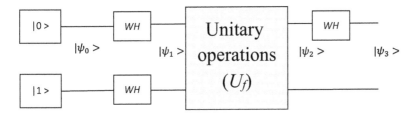

FIGURE 1.2   Block diagram of Deutsch's algorithm

gate in quantum computation and plays a significant role in Deutsch's algorithm. Consider Figure 1.2 for Deutsch's algorithm.

The input state, expressed as $|\psi_0\rangle = |0\ 1\rangle$, is passed through two *WH* gates, which results in $|\psi_1\rangle$.

$$
|\psi_1\rangle = \begin{bmatrix} \dfrac{|0\rangle + |1\rangle}{\sqrt{2}} \\[2mm] \dfrac{|0\rangle - |1\rangle}{\sqrt{2}} \end{bmatrix}
\tag{1.6}
$$

The obtained results of $|\psi_1\rangle$ then undergo a certain unitary transformation (U_f) depending on the function $f(x)$ based on the considered problem to be solved. It then provides two possibilities, for example:

$$
|\psi_2\rangle = \begin{bmatrix} \pm\left[\dfrac{|0\rangle + |1\rangle}{\sqrt{2}}\right]\left[\dfrac{|0\rangle - |1\rangle}{\sqrt{2}}\right] iff\,(0) = f(1) \\[3mm] \pm\left[\dfrac{|0\rangle - |1\rangle}{\sqrt{2}}\right]\left[\dfrac{|0\rangle - |1\rangle}{\sqrt{2}}\right] iff\,(0) \neq f(1) \end{bmatrix}
\tag{1.7}
$$

The final *WH* gate on the first qubit thus gives us

$$
|\psi_3\rangle = \begin{bmatrix} \pm|0\rangle\left[\dfrac{|0\rangle - |1\rangle}{\sqrt{2}}\right] iff\,(0) = f(1) \\[3mm] \pm|1\rangle\left[\dfrac{|0\rangle - |1\rangle}{\sqrt{2}}\right] iff\,(0) \neq f(1) \end{bmatrix}
\tag{1.8}
$$

The above solution can be combined as below based on the two possibilities:

$$
|\psi_3\rangle = \pm\,|f(0) \oplus f(1)\rangle\left[\dfrac{|0\rangle - |1\rangle}{\sqrt{2}}\right]
\tag{1.9}
$$

where $f(0) \oplus f(1)$ may be measured by measuring the first qubit and $f(0) \oplus f(1) = 0$ *iff* $(0) = f(1)$ and 1, otherwise. This is quite an interesting result. It indicates how the quantum circuit can determine a global property of $f(x)$, namely

$f(0) \oplus f(1)$, with only one evaluation of $f(x)$. This is quicker than any classical computing machine, which needs at least two assessments.

1.6 QUANTUM COMPUTATION IN SATELLITE COMMUNICATIONS

Due to the rapid development of science and technology in recent years, the performance of communication equipment has improved in terms of efficiency, speed and latency. These technologies have further been enabling secure and reliable communication using quantum science because conventional communication methods possess a mutual restriction between complexity and communication performance. Based on the principle of quantum superposition effect and quantum entanglement, a quantum communication system or quantum network has now emerged as a next-generation communication technique for inner and outer space of the earth that provides low latency, high reliability and constant data flow, even in a very dense network. A quantum network is made up of many distinct nodes, each of which stores quantum information. It also allows transfer of information in the form of qubits between physically distant quantum processors, each of which is a miniature quantum computer capable of performing quantum logic gates (operation) on a set of number of qubits [37]. Primarily, there are four types of free-space satellite communication as mentioned in Table 1.2.

Successful free-space communication is not easy to achieve. Background photons acquired by the satellite, on the other hand, would be the source of the errors. Whether the moon is full or new, it influences the background rate; during full moons, background photons dominate the error rate, but detector noise dominates during new moons. During daytime orbits, the background radiance would be far greater. It is believed that quantum computing based satellites would be useful to provide more secure, safe and successful free-space communications. Following the secret Quantum Key Distribution (QKD) protocol, a quantum satellite can have entangled distribution across vast distances [39]. As a result, it would be possible to

---

TABLE 1.2
Types of Satellite Communications

S. No.	Type of Communication	Working Range	Frequency Handling Range	Remark
1	Open-air	Up to 100 km above the earth's surface	~ 100–1,000 kHz	Horizontal/local communication
2	Earth-Satellite	Between 300 km and 800 km above the earth's surface	~ 0.1–10 MHz	Drone and small satellite to earth
3	Satellite broadcast	Earth's orbit ~ 36,000 km	~27 MHz	Broadcasting of channels, GPS, etc.
4	Inter-satellite	Earth's orbit	Any frequency	Free-space communications between satellite(s)

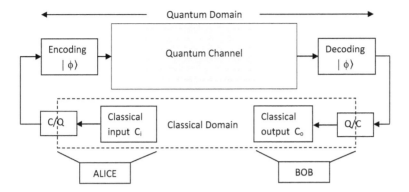

FIGURE 1.3 Classical information (Ci) transmission through quantum channel of quantum satellite.

send encrypted information using all types of satellite communication systems. The inherent QKD algorithm in the quantum communication through a quantum satellite may be effective in securing data distributed throughout the world and provides an un-hackable, loss-free and distortion-free communication.

1.6.1 WORKING OF A TYPICAL QUANTUM-BASED SATELLITE SYSTEM

The communication of classical information from sender to receiver end via a quantum satellite is depicted in Figure 1.3. The communication between sender and receiver can be listed in the following steps.

Step-1: Classical input (C_i) at the sending side (Alice) is transformed into quantum input and encoded as $\phi >$.

Step-2: The encoded quantum signals in the form of quantum states are sent through a quantum channel. Interaction with the environment introduces noise which is in a quantum sense 'uncertainty'. It can be remembered that the errors in a quantum channel are different from the errors that may arise in classical channels [37].

Step-3: The received signal at the receiving end (Bob) is decoded by quantum transformations on qubit.

Step-4: The actual classical measurements give classical output [40].

Successful experimentations make free-space networks realistic today. 'Micius' satellite, made in China, was used to realize a ground space free communication network [41]. However, further exhaustive research can only confirm better performance in near future.

1.7 CONCLUSIONS AND FUTURE SCOPE

We hope that the matter discussed in the present chapter will be helpful to the readers to introduce them to the understanding of quantum information processes and the role of quantum logic gates. The scope and depth of literature presented in this chapter in some places might appear to be limited, but that is deliberate. The concepts presented for, say logic gates, are the simplest and are chosen for the introduction to quantum computation processes. The complex and more advanced logic gates and their processes applicable for quantum computers have been omitted here, as the objective is not to straight away give the expertise, but to build a basic understanding in the subject, which can further be developed. Hence, a basic idea of how quantum computation is envisaged to be utilized in satellite communication is presented in this chapter. In addition, to complete the picture, a schematic of the working of a quantum satellite communication system is presented here. The readers after going through the present chapter, hopefully, will acquire the basic ideas required to go for advanced topics of quantum computations involved in drone and satellite communication processes. Also, we believe that the readers would find some of the topics, like quantum entanglement, attractive due to the possibility of teleportation applicable to the transportation of information over long distances, and for its potential applications in drones and satellite communications.

REFERENCES

[1] A. Barenco, C. H. Bennett, R. Cleve, D. P. DiVincenzo, N. Margolus, P. Shor, T. Sleator, J. A. Smolin, and H. Weinfurter (1995). Elementary gates for quantum computation. *Phys. Rev. A*, 52, 3457.

[2] P. W. Shor (1994). Algorithms for quantum computation: discrete logarithms and factoring. In: *Proc. 35th Annual Symp. on Found. of Computer Science.* IEEE Computer Society, Los Alamitos, pp. 124–134.

[3] D. Deutsch (1985). Quantum theory, the Church-Turing principle and the universal quantum computer. *Proc. Roy. Soc. Lond. A*, 400, 97–117.

[4] A. Saharia, R. K. Maddila, J. Ali, P. Yupapin, and G. Singh (2019). An elementary optical logic circuit for quantum computing: a review. *Opt. Quantum Electron.*, 51, 1–13.

[5] T. B. Pittman, B. C. Jacobs, and J. D. Franson (2001). Probabilistic quantum logic operations using polarizing beam splitters. *Phys. Rev. A*, 64, 062311.

[6] E. Knill, R. LaFlamme, and G. J. Milburn (2001). A scheme for efficient quantum computation with linear optics. *Nature*, 409, 46–52.

[7] M. Koashi, T. Yamamoto, and N. Imoto (2001). Probabilistic manipulation of entangled photons. *Phys. Rev. A*, 63, 030301.

[8] S. Gasparoni, J. -W. Pan, P. Walther, T. Rudolph, and A. Zeilinger (2004). Realization of a photonic CNOT gate sufficient for quantum computation. *Phys. Rev. Lett.*, 93, 020504.

[9] M. A. Nielsen (2005). Optical quantum computation using cluster states. *Phys. Rev. Lett.*, 93, 040503.

[10] R. Raussendorf, and H. J. Briegel (2001). A one-way quantum computer. *Phys. Rev. Lett.*, 86, 5188–5191.

[11] Z. Zhao, A. -N. Zhang, Y. -A. Chen, H. Zhang, J. -F. Du, T. Yang, and J. -W. Pan (2005). Experimental demonstration of a nondestructive controlled-NOT quantum gate for two, independent photon qubits. *Phys. Rev. Lett.*, 94, 030501.

[12] N. K. Langford, T. J. Weinhold, J. L. O'Brien, R. Prevedel, G. J. Pryde, K. J. Resch, and A. G. White (2005). Demonstration of a simple entangling optical gate and its use in Bell-state analysis. *Phys. Rev. Lett.*, 95, 210504.

[13] T. P. Spiller, W. J. Munro, S. D. Barrett, and P. Kok (2005). An introduction to quantum information processing: applications and realizations. *Contemp. Phys.*, 46, 407–436.

[14] P. Kok, and S. L. Braunstein (2006). Relativistic quantum information processing, with bosonic and fermionic interferometers. *Int. J. Quantum Inf.*, 4, 119–130.

[15] P. Kok, W. J. Munro, K. Nemoto, and T. C. Ralph (2007). Linear optical quantum computing with photonic qubits. *Rev. Mod. Phys.*, 79, 135–174.

[16] X. -D. Cai, C. Weedbrook, Z. -E. Su, M. -C. Chen, M. Gu, M. -J. Zhu, L. Li, N. -L. Liu, C. -Y. Lu, and J. -W. Pan (2013). Experimental quantum computing to solve systems of linear equations. *Phys. Rev. Lett.*, 110, 230501.

[17] S. Barz, I. Kassal, M. Ringbauer, Y. O. Lipp, B. Dakić, A. Aspuru-Guzik, and P. Walther (2014). A two-qubit photonic quantum processor and its application to solving systems of linear equations. *Sci. Rep.*, 4, 6115.

[18] A. Carolan, C. Harrold, C. Sparrow, E. Martín-López, N. J. Russell, J. W. Silverstone, P. J. Shadbolt, N. Matsuda, M. Oguma, M. Itoh, G. D. Marshall, M. G. Thompson, J. C. F. Matthews, T. Hashimoto, J. L. O'Brien, and A. Laing (2015). Universal linear optics. *Science*, 349, 711–716.

[19] T. Meany, M. J. Steel, M. Grafe, M. J. Withford, R. Heilmann, P. -L. Armando, and A. Szameit (2015). Laser written circuits for quantum photonics. *Laser Photon. Rev.*, 9, 363–384.

[20] J. Zeuner, A. N. Sharma, M. Tillmann, R. Heilmann, M. Gräfe, A. Moqanaki, A. Szameit, and P. Walther (2018). Integrated-optics heralded controlled-NOT gate for polarization-encoded qubits. *NPJ Quantum Inf.*, 4, 13–20.

[21] Ben Ahmed, A., and A. Ben Abdallah (2016) An energy-efficient high-throughput mesh-based photonic on-chip interconnect for many-core systems. *Photonics*, 3(2), 15–38.

[22] J. Wang, S. Paesani, R. Santagati, S. Knauer, A. A. Gentile, N. Wiebe, M. Petruzzella, J. L. O'Brien, J. G. Rarity, A. Laing, and M. G. Thompson (2017). Experimental quantum Hamiltonian learning. *Nat. Phys.*, 13, 551–555.

[23] V. M. Schäfer, C. J. Ballance, K. Thirumalai, L. J. Stephenson, T. G. Ballance, A. M. Steane, and D. M. Lucas (2018). Fast quantum logic gates with trapped-ion qubits. *Nature*, 555, 75–78.

[24] H. L. Huang, X. -L.Wang, P. P. Rohde, Y. -H. Luo, Y. -W. Zhao, C. Liu, L. Li, N. L. Liu, C. Y. Lu, and J. -W. Pan (2018). Demonstration of topological data analysis on a quantum processor. *Optica*, 5, 193–198.

[25] H. -L. Huang, Q. Zhao, X. Ma, C. Liu, Z. -E. Su, X. -L. Wang, L. Li, N. -L. Liu, B. C. Sanders, C. -Y. Lu, and J. -W. Pan (2017). Experimental blind quantum computing for a classical client. *Phys. Rev. Lett.,* 119, 050503.

[26] C. Liu, H. -L. Huang, C. Chen, B. -Y. Wang, X. -L. Wang, T. Yang, L. Li, N. -L. Liu, J. P. Dowling, T. Byrnes, C. -Y. Lu, and J. -W. Pan (2019). Demonstration of topologically path-independent anyonic braiding in a nine-qubit planar code. *Optica*, 6, 264–268.

[27] A. Kumar, S. Bhatia, K. Kaushik, S. M. Gandhi, S. G. Devi, A. Diego, D. J. Pacheco, and A. Mashat (2021). Survey of promising technologies for quantum drones and networks. *IEEE Access*, 9, 125868–125911.

[28] R. Barends, A. Shabani, L. Lamata, J. Kelly, A. Mezzacapo, U. L. Heras, R. Babbush, A. G. Fowler, B. Campbell, Y. Chen, and Z. Chen. (2016). Digitized adiabatic quantum computing with a superconducting circuit. *Nature*, 534, 222–226.

[29] A. Turing (1936). On computable numbers with an application to the Entscheidungs-problem. *Proc. Lond. Math. Soc.*, S2–42, 230–265.

[30] A. Church (1936). An unsolvable problem of elementary number theory. *Am. J. Math*, 58, 345–363.

[31] D. Bruss (2013). Quantum information processing. Lecture notes. In: Chapter A2 of *Introduction to Quantum Information*. ForschungszentrumJulich GmbH.

[32] D. Deutsch (1989). Quantum computational networks. *Proc. Roy. Soc. Lond. A*, 425, 73–90.

[33] D. P. DiVincenzo (1995). Two-bit gates are universal for quantum computation. *Phys. Rev. A*, 51, 1015–1018.

[34] T. Sleator, and H. Weinfurter (1995). Realizable universal quantum logic gates. *Phys. Rev. Lett.*, 74, 4087–4090.

[35] A. Barenco (1995). A universal two-bit gate for quantum computation. *Proc. R. Soc. Lond. A*, 449, 679–683.

[36] A. Barenco, C. H. Bennett, R. Cleve, D. P. DiVincenzo, N. Margolus, P. Shor, T. Sleator, J.A. Smolin, and H. Weinfurter (1995). Elementary gates for quantum computation. *Phys. Rev. A*, 52, 3457–3467.

[37] J. Chen (2021). Review on quantum communication and quantum computation. *J. Phys.: Conf. Ser.*, 1865, 022008.

[38] M. A. Nielsen and I. Chuang (2010). *Quantum Computation and Quantum Information*. 10th Anniversary Edition, Cambridge University Press, Cambridge.

[39] J. Yin, Y. Cao, Y.H. Li, S.K. Liao, L. Zhang, J.G. Ren, W.Q. Cai, W.Y. Liu, B. Li, H. Dai, and G.B. Li (2017). Satellite-based entanglement distribution over 1200 kilometers. *Science*, 356, 1140–1144.

[40] L. Hanzo, H. Haas, S. Imre, D. O'Brien, M. Rupp, and L. Gyongyosi (2012). Wireless myths, realities, and futures: from 3G/4G to optical and quantum wireless. *Proc. IEEE*, 100, 1853–1888.

[41] Z. Zhihao (2017). Beijing-shanghai quantum link a 'new era'. China Daily, 2017–2019.

2 A Brief Study on Quantum Walks and Quantum Mechanics

Sapna Renukaradhya, Preethi, Rupam Bhagawati, and Thiruselvan Subramanian
Presidency University

CONTENTS

2.1 QUANTUM

People have been perplexed by the laws of nature's counter-intuitive nature since the discovery of quantum physics. Over time, we've come to accept a growing number of consequences that would be unthinkable in a Newtonian universe. Modern

DOI: 10.1201/9781003250357-2

technology makes use of quantum effects, both, for our gain and for our detriment, as some of the most outstanding ones are laser technology and the atomic bomb. Quantum information theory arose from the desire to use its laws to create devices with incredible capability, such as quantum cryptography and quantum computation [1]. In 1994, Peter Shor found a quantum algorithm for factorizing numbers effectively, igniting a spike in interest in physics, computer science, and computation. This quantum factorisation approach is significantly faster than any traditional method [2]. While no general-purpose quantum computer has yet been developed, quantum algorithm studies are indeed ongoing for many decades.

This promising research line has found a plethora of new phenomena that are remarkably distinct from their traditional equivalents, both physically and in terms of computer science and communication theory. These groups have gotten a better comprehension of one another's concepts and ideas over time. The notion that information cannot be isolated from the physical device that carries it, as per the statement "Information is physical", has taken root among researchers, leading to some exciting scientific breakthroughs. Understanding the fundamental concepts of all of these domains appears to be critical to comprehending current quantum information processing.

Modern machines are now more powerful, although there are still severe restrictions to the complexity that can be addressed and the amount of time it takes to solve them. Quantum computers, which use quantum mechanical principles to encode data into quantum particles known as qubits and execute computations, can break a few of these constraints. In the topic of quantum algorithms, physicists and computer scientists look at traditional computing algorithms and concepts to determine whether there was a more efficient or faster technique that can be implemented on a quantum computer, known as a "quantum advantage".

The work carried out by Y. Aharonov et al. [3] in 1993 can be considered a forerunner of later models. For the first time, their study coined the word "quantum". We can use quantum physics to post-select events that can't be seen in a traditional way. Since these are rare occurrences, there is no traditional setup that permits us to observe such an impact. The "Gedankenexperiment" reveals some of the quantum weirdness.

The discussion in the chapter is organised into sections like Quantum, Computing, Random Walks, The Quantum Walk, Quantum Walk Principle, The Quantum Random Walk Search, Random Walk in Computer Science, Quantum Computing with Quantum Walks and Related Models, Hybrid Classical-Quantum Algorithms, Quantum Mechanics, Relationships between Quantum Walk and Quantum Mechanics, Quantum Computers and Circuits, Drone-Based Quantum Computing, and Quantum Satellites for Drone-Based Network and Communications.

2.2 COMPUTING

In mathematics, computation is defined as a process on input data. While no clear and concise definition exists, fuzzy definitions are suitable for this work, which is concerned with the potential and limits that the laws of physics impose on computation. A computer can be defined as a physical system that performs a specific computation.

A mobile phone, your intellect, and an abacus, for instance, are all physical systems that may be computers. Until around 1950, the term "computer" referred to individuals performing computations that remained as components of a bigger complete calculation and may be completed quicker in parallel [4]. It can be immediately declared that properly carrying out a real computation must generate an output for all of these various machines. A brick doing nothing can be used to substitute any computation with no output. This effectively eliminates pan-computationalism, which holds that everything in the cosmos is a computer [4]. This is not a good viewpoint to hold since then there's no difference between a computer and any other form of a physical system. Horsman et al. propose a more precise explanation and structure in their work [5].

Now let us discuss efficient computation. Computing is a high-level process in which manufactured things and computational entities interact. Because a computational entity is required, computing must be having a purpose or some use. This means that computation is done so that it solves an issue as quickly as possible. Now we will look for what is required for efficient computing and what enables one computer to perform at a higher level than another for a given task. This subject has many aspects, and we will only look at a few essential refrains that are relevant to quantum computing.

We must not forget that efficiency has varied meanings for diverse purposes. On the one hand, complexity experts think about computations scaling asymptotically. To them, efficiency means a polynomial scaling of computer resources in proportion to the magnitude of the issue. For real computation, on the other hand, efficiency includes acquiring answers on the timeframes appropriate to the computational entity. The timescale for producing visuals is quicker than even the eye can perceive flickers. Days or weeks are feasible for challenging simulations. It signifies that it is more rapid than the physical process simulated for real-time tracking or prediction; a flawless weather forecast is not as useful just after the weather has transpired. When it comes to designing a viable quantum computer, the intricacy of a task and the expected quantum benefit are solid beginning points, but not the full story.

A better prefactor in the scaling might be sufficient to justify the use of a quantum computer, particularly for time-sensitive computations. To obtain a benefit, it is not essential to tackle a greater issue than is traditionally doable. Based on the price of the answer, resolving a problematic scenario rapidly is sufficient to obtain a reasonable edge.

The efficient encoding of the data is a second key consideration. One of the reasons why traditional digital computers operate so well is because of binary encoding. Consider an instance of in what way binary encoding is exponentially more memory efficient than unary encoding. As the amount of people grows, this becomes more crucial. Digitalised computers employ even more efficient representation, such as floating-point numbers, which strike a balance amid accuracy in addition to control. Same encoding considerations are applied during the development of quantum computers [6]. Furthermore, since the measurement process in quantum physics is non-trivial and therefore can change the state of the registers, the encoding affects the measurement process's efficiency and accuracy. The aforementioned is important to amount in a method that can differentiate among N alternative consequences for a unary encoding. As N grows larger, this gets incredibly hard. To discriminate N numbers in binary encoding, $\log_2 N$ amounts having two results each are sufficient. Because each measurement only has to differentiate between two orthogonal results, it is more precise. Twice the problem's size (N) just necessitates another binary measurement, which is additional effectual when compared to twice the quantity of measurement consequences to discriminate. When it comes to theoretical computer science in its canonical form, the physical features of machines that are utilised to conduct computational or data processing processes are not taken into consideration. Since the behaviour of any physical equipment used for computation or data processing must ultimately be predicted by physical principles, several research approaches have been established to conceptualise about computing within a physics framework. When it comes to physical theories that may be employed for this purpose, quantum mechanics is the most ancient and sophisticated of the ones available.

In quantum computing, it is an integrated scientific discipline that is devoted to the creation of quantum computers and quantum data processing systems that use the quantum mechanical properties of nature. Quantum computing is a scientific area that is separated into two parts: quantum computation and quantum data processing. Quantum computation is a subfield of quantum data processing. While studying quantum computing, researchers are interested in the creation and operation of algorithmic techniques which use the physical characteristics of quantum computers. Developing efficient quantum algorithms because of the paradoxical nature of quantum physics and the fact that intuition plays an essential role in algorithm design makes this task tough to do successfully. An algorithm based on quantum mechanics is needed to perform functions beyond just accomplishing the task for which it was designed. In order to be regarded as superior to another classical algorithm, it must outperform or be more efficient than the other algorithm [7].

A number of investigations have yielded successful examples of quantum computing results in the real world. A particularly successful technique for developing quantum algorithms may be to use quantum walks, which are the quantum mechanical equivalents of conventional random walks. Quantum walks are the quantum mechanical analogues of ordinary random walks. Recently in Ref. [8], it was shown that there is a universal quantum computing paradigm, which is now generally recognised.

2.3   RANDOM WALKS

There have been significant advancements in the disciplines of computer science and computer engineering, and these advancements have had a significant influence on almost every element of contemporary life. The development of new models of computation, the development of new materials and techniques for the construction of computer hardware, the development of novel methods for speeding up algorithms, and the establishment of bridges between computer science and a variety of other scientific fields are all examples of innovative research in these fields. In this way, scientists may conceive of natural occurrences as computing operations, while simultaneously simulating them using fresh computational models. Our top priority should be quantifying the resources required to process data and calculate a solution, which is another way of saying that we should evaluate the difficulty of a computing process, because it allows us to assess implementation costs and evaluate problems by evaluating the overall intricacy of their solutions, which is a high-priority research field. Besides being extensively employed in many fields of knowledge, such as physics, biology, commerce, computer vision, and earthquake modelling, classical random walks also play a significant role in algorithm design and implementation, as well as in algorithm design and implementation [9] [10].

It is referred to as a random walk when an unpredictable approach is used to design a path in a mathematical space that includes a succession of randomly chosen steps. It is becoming more used in a variety of fields, including mathematical computations and computer science, among others. The random walk is a computing paradigm that is widely used and well sophisticated in a variety of fields, including mathematics, natural science, and computer science. Classical random walks, a subset of stochastic processes (practices whose growth is based on chance), have proven to be a particularly effective tool for the construction of stochastic algorithms among the mathematical tools used in advanced computation, particularly in the field of advanced computing. Classical random walks are extensively employed in many fields of knowledge, including physics, natural science, finance theory, computer vision, and catastrophe prediction. They also play an important role in algorithmic design, which is why they are so often used. Physical properties of systems that are utilised to carry out computation or information processing activities are not taken into consideration at all when it comes to computational linguistics in its canonical form. In order for the behaviour of whatever physical systems are used for computing or data processing to be ultimately predicted by physics principles, various research approaches have been established that are focused on investigating computation within a physical framework.

The quantum walk is a quantum counterpart of the conventional random walk that is used in quantum computing (also known as quantum walk). In comparison to regular computers, quantum walk, which takes advantage of the phenomenon of quantum superposition [9], provides an exponential algorithmic speedup. Despite the fact that many years have passed and various improvements have been made, and some of the unresolved problems have been thoroughly explored and answered [10,11], further study is still required for proper examination of the quantum walk phenomenon. Search results for Google's search engine may be found by using the PageRank algorithm [12], which performs a random walk through a network with vertices representing different websites to identify relevant results. Rather than just exploring the internet at random, it uses this method to evaluate the relevance of every given piece of information. According to research, the PageRank algorithm has been customised in a number of ways, including the use of personalised PageRank [13,14]. Other novel algorithms, such as Random Walk with Restart (RWR) [15] and Lazy Random Walk (LRW) [16], are proposed in addition to walk rules.

Classical random walks, as well as quantum random walks, may be used to compute the distance between nodes and to determine the topology of a network, among other applications. Random walk-related models may be used in a variety of domains and are particularly beneficial for downstream tasks such as link prediction, recommendation, computer vision, semi-supervised learning, and network embedding, to mention a few examples.

Random walks are often used in computer science and other domains to replicate natural phenomena. This is a frequent practice in computer science and other fields. Traditionally, when it comes to computers, they are represented as a series of random steps on a mathematical space. Random walks are a type of stochastic model (i.e., a random mathematical process) defined as Markov chains, which is a process in which forecasts can be produced with the same confidence at any stage in the process despite knowing the previous states [17]. The large variety of classical applications for random walks, including reinforcement learning, random number generation, and thermodynamics, leads one to ask if random walks realised with quantum computers have a similar range of uses. According to recent research [18], there is a quantum advantage in both continuous- and discrete-time quantum random walks, as well as classical random walks with discrete state-space. A considerable variation in behaviour may also be seen between the two groups.

Once we sample out of a set of objects per some distribution, we could come up with an answer to many computational problems [19]. Building a random walk on a graph in which nodes are the objects to be sampled, is a common way of approaching this challenge. The graph beside the walk is established so that the limiting distribution is the same as the one we want to sample from.

For sample, if one begins the walk at a random starting position and can watch it progress. Unless the random walk reaches the limiting distribution quickly, this type of technique is efficient. The degree of convergence to the limiting distribution can be defined in a variety of ways.

2.4 THE QUANTUM WALK

Quasi-classical random walks, which are also known as quantum analogues of quantum walks in other domains such as mathematics and physics, are analogous to classical random walks in quantum mechanics. Quantum random walks are based on the notion of iterating the walk without the need for intermediate measurements, and they may be thought of as a kind of game in and of itself. Quantum random walks are a type of game in and of themselves. Alternatively, we shall proceed with the method of unitary translation and rotation without pausing to collect any measurements at any of the intermediate time periods in either direction.

As part of this technique, we will discretise the particle's position space, which implies that as a consequence of this operation, the particle will be placed on a lattice or graph. This site will be closed off to the general public for most of the time. Bounded state-space condition is required in the context of simulability on a confined computer because a discrete and limited state space is required for simulability on a constrained computer. With the usage of discrete registers in a quantum computer, its state space is a huge but confined Hilbert space. It is possible to convert the quantum walk into a calculation that can be carried out by this kind of computer if we discretise the quantum walk.

As a result of studying quantum random walks, we will learn about some exciting new physics, and we will be able to leverage the "abnormal" parts of the walk that are made possible by quantum mechanics to increase our computing capacity. Performing a quantum random walk and using the results to successfully accomplish tasks that have been allocated to the quantum computer is something that this computer is capable of doing. Beyond its use in computer science, the notion of quantum walks may be used to establish methods of verifying the "quantumness" of current technologies, as well as to construct quantum computers and replicate natural phenomena.

Quantum walks are well-known for their mathematical complexity and wide range of applications. There is a significant distinction between using quantum walks for computation and using them to simulate physical systems. The purpose of computing is to create an efficient algorithm, while the goal of modelling is to properly characterise the physical properties of the system. Physical quantum walks have been handled in most quantum walk experiments [20], with a walker like photon or atom, travelling a path in the experimental setup as the quantum walk progresses. There are some algorithmic applications of quantum walks, such as [21], in which the walk is represented by qubits, that mark the walker's position.

Take, for example, random walks, which can be performed effectively by a quantum computer if the equivalent classical random walk can be implemented successfully by a classical computer. When a conventional circuit that performs the classical random walk is converted into a quantum circuit (after it has been modified to make it reversible), and when a conventional coin flip is replaced with a quantum coin flip, which can be performed effectively on the limited space available for coin-qubits, this is known as a quantum coin flip.

As an alternative expression, if we start with the classical random walk as the basis for a classical algorithm and show that the quantum random walk improves its behaviour, we will have a quantum algorithm that is much quicker than the classical method. When it comes to the continuous-time quantum random walks model, the picture isn't quite as clear as it seems. Although this is true, it is not immediately obvious how to adapt its continuous development to the discrete quantum circuit model from the commencement of the simulation. It has previously been shown [3] that a quantum computer is capable of generating continuous-time random walks with high precision in a controlled setting.

New and intriguing research topics in the realm of information technology include the investigation of connections between classical and quantum random walks, as well as the application of quantum walks to computer science problems. The scientific evidence on the linkages between classical and quantum travels, as well as the strong mathematical connectivity between correlated random walks and quantum walks, which is produced by using the PQRS matrix technique [22], demonstrates that quantum walks outperform classical walks in terms of efficiency and effectiveness. The following are the specifics of each of the two quantum walk models that have been suggested so far.

2.4.1 The Discrete Model

Starting with the basic setup for the first model, the discrete-time quantum random walk, which will be explored later, describes the model's fundamental setup. A walk over a number line, which is a well-known example of a discrete random walk, is a good example of a discrete random walk. In the case of a number line and a coin, the position of one on the number line will change one step to the left or right depending on which way the coin is flipped. Throughout the walk, there is a good possibility that the next location will be the previous position or the next integer on the number line, and transition probabilities are solely dependent on the step in which the transition takes place.

This model is made up of two quantum mechanical systems, which are represented by the walking particle and the coin, as well as an evolution operator that is applied to both systems predominantly at discrete time intervals. The mathematical structure of this model entails development via the use of a unitary operator, which is similar to that used in Eq. (2.1).

$$|\psi >_{t_2} = \hat{U} |\psi >_{t_1} \qquad (2.1)$$

As a consequence of this investigation, Feynman's writings [23] take into consideration the discretisation of the Dirac equation, and a model is developed as a result of this investigation. It has been tried to be recreated in the setting of the quantum Turing computer stopping by many writers, including Meyer [24], Watrous, and others [25,26]. Also in Ref. [25], reasoning about space-bounded quantum computing techniques, where it was connected with quantum cellular automata, it was replicated with a little alteration using measurement methods, and it was also linked with quantum cellular automata. Ambainis et al. [26] created and tested it as a practical computing tool that conformed to a formal presenting style, which they called "formal presentation style".

2.4.2 CONTINUOUS-TIME QUANTUM RANDOM WALK

Interspersing quantum phenomena with random walks is possible when employing quantum random walks in discrete time, which is one means of doing this. Another use of this principle is continuous random walks, the distinction being that they do not need the additional dimension of a coin space in order for them to work efficiently and effectively. The evolution (Hamiltonian) operator should be utilised for continuous quantum walks because it is capable of being employed without any restrictions, which implies that the walker may walk at any point in the system's evolutionary history without being restricted. Construction of the mathematical basis for this model is based on the Schrodinger equation, which portrays the process of evolution. Continuous-time in addition to creating discrete state spaces for each state in the state space, a transition rate matrix is also created for each state in the state space, which provides the "jumping rate" or likelihood of transition between any two adjacent states in the state space for every state in the state space. It is feasible to build the walk as a decision tree using the techniques described in Ref. [27], and a quantum operator can be created to traverse it using the methods presented in Ref. [27].

Discrete graphs, as well as other topologies, were used to execute quantum walks and to estimate the properties of quantum walks in both discrete and continuous models, as was the case with the previous work. In part, this is due to the fact that graphs are extensively used in computer science and that the development of quantum algorithms based on quantum walks has been a major emphasis over the last decade. A broad number of physical processes have been represented using quantum walk models, and these models have also been employed as an algorithmic tool in the development of quantum computing. As shown by the quantum walk model, quantum particles may be transported in a coherent fashion across a range of diverse surroundings without losing their properties. This adaptive notion may be used in a wide range of systems, including physical, chemical, and biological systems, in order to characterise quantum particle dynamics. Furthermore, quantum walks provide a fresh and intriguing paradigm for the use of quantum computing technologies. There are a variety of optical designs that make use of quantum walks, ranging from fibre-based experiments to integrated waveguide arrays, and they are all important in their own manner.

2.5 QUANTUM WALK PRINCIPLE

The adaptability of its classical equivalent, the classical random walk, inspires quantum walk study. Random walks are used to explain scientific phenomena such as Brownian motion, as a foundation for search engines, and even to forecast the price progression on the stock market. Research is increasingly focusing on algorithms that depend on the profit from the underpinning quantum mechanics in quantum walks, as a result of the varied applications. Coherent mechanisms in nature, such as energy flow in photosynthesis, are a good example.

In a traditional random walk, a walker chooses his path at random may be by tossing a coin, depicted as a simpler Galton board experimentation. A binomial distribution of a range proportional to the square root of steps gives the likelihood of ending up at a specified final position after a discrete number of steps (rows). A quantum walker, on the other hand, is characterised by a quantum mechanical wave function. The wave function extends across numerous locations through evolution, moving out of the arrangement in a coherent superposition of all conceivable trajectories. Quantum interference effects emerge, substantially altering the ultimate probability distribution. A quadratically faster propagation in a quantum walk reduces the overall path to catch distant sites, which is a major difference between the two models. In contrast to the classical random walk, the growth of a quantum walk is totally deterministic, with the ultimate distribution strongly dependent on the ambient factors.

2.6 THE QUANTUM RANDOM WALK SEARCH

The discrete quantum random walk technique around a hypercube [28], which is a search strategy that is based on a discrete quantum random walk algorithm, has been given, despite the fact that quantum random walk algorithms are still in their infancy. In terms of both structure and function, the quantum random walk search algorithmic steps and unitary operator seem to be structurally and functionally analogous to Grover's search techniques in both structure and function. When compared to other quantum search algorithms, the random walk's distinguishing characteristics make it simpler to develop but more difficult to investigate.

Grover's search method and the quantum random walk search algorithm have certain characteristics in common, and the two algorithms are similar in other ways. In order to fully study and grasp Grover's implementation mechanics, it is necessary to first thoroughly investigate and understand the quantum random walk search. It is highly recommended that you complete this step before proceeding. A two-dimensional rotation is illustrated in this picture as a consequence of the fact that Grover's approach starts in a superposition of all possible states, includes a "marking" step, and then applies the Grover diffusion operator. Grover's approach is based on the principle of superposition, which means that all conceivable states may exist simultaneously. In addition, [28] points out that there are some important differences between the two strategies, which will be described in further depth lower down the page.

It is not suggested to use the random walk search when dealing with a two-dimensional subspace since it does not map an accurate two-dimensional subspace. For example, whereas the subspace of the Grover walk is entirely covered by a superposition of all states plus the target state, the subspace of the walk is completely covered by a superposition of all states plus an approximation to the goal state for the walk. When compared to Grover's approach, in which the end state is equal to the target state, the final state receives just a little amount of contribution from its neighbour nodes, which is a significant difference. It has not yet been determined if these inconsistencies are advantages or disadvantages, and the matter will remain unresolved until further research has been undertaken to assess these algorithms on a range of quantum computing hardware components.

2.7 RANDOM WALK IN COMPUTER SCIENCE

For those unfamiliar with theoretical computer science, random walks are best stated as follows: In addition to estimating the bulk of convex bodies, approximating the permanence of matrices, and identifying tasks aimed at Boolean equations as algorithmic tools, they have been used to a variety of other issues. A series of basic, local adjustments is used by the researchers to create a universal paradigm for collecting and assessing an ever-increasing variety of combinatorial objects [19].

Several random walk algorithms, such as the Markov Chain Monte Carlo Method [19], are influenced by factors such as the walk's mixing time (the time it takes for the walk to reach its static distribution; this is necessary if we want to sample from a set of combinatorial structures according to a distribution; see the Markov Chain Monte Carlo Method [19]), or the number of expected hitting instances that exist between the two vertices of the implicit graph [20].

2.8 QUANTUM COMPUTING WITH QUANTUM
WALKS AND RELATED MODELS

Childs et al. [9], for example, established the quantum walks as universal for quantum computing. This solution for a single walker, on the other hand, only applies to quantum walks coded on qubits. The quantum walks are inefficient without this encoding in terms of complexity theory growing with issue size, and hence will not give a quantum benefit.

A second finding from Childs et al. [9] demonstrates that many interactive quantum walkers give universal quantum computing without the need for qubit encoding.

Physical quantum walks are efficient in this scenario and are equal to quantum cellular automata [29]. Boson sampling [30], which consists of many non-interacting quantum walkers, is midway between the both. Although boson sampling does not necessitate qubit encoding, it is not universal in quantum computing. It can still perform some difficulties more quickly than the famous classical technique, like estimating the permanence of a difficult matrix.

Concerning the same issue, there are various quantum walk algorithms through intricacy proofs of their speedup above classical techniques. The first such method [8] searches an unsorted database. Quantum walk search has become a workhorse sub-function for researchers building algorithms for more difficult problems, delivering a quadratic speedup over classical algorithms. Random guessing is the optimal classical technique for an unstructured database, and it grows linearly with problem size N, whereas quantum search scales as $\sqrt{N}$.

The "glued trees" technique [9], when compared to an oracle, gives an exponential speedup, but it does not contribute to solutions in the same way that quantum search does [9,10]. In the course of applying the search method or the glued trees methodology, it is possible to encounter permutation symmetric issues. When employing testbed approaches, the performance of quantum hardware may be tested using permutation symmetric tasks, which is especially beneficial for analysing the performance of quantum hardware. When difficulties can be solved analytically, effectively calculated conventionally (as long as the findings are known), and successfully integrated into quantum computers, a comprehensive evaluation of the efficiency of quantum devices is conceivable. It is conceivable to employ extra qubits as gadgets in order to establish permutation symmetry, which may then be used to correctly execute on quantum hardware [31], which is currently being investigated.

A random walk is constructed using the network topology, which may also be used to determine the distance between nodes as a result of its usage in the development of random walks. In the field of collaborative filtering, researchers have looked at algorithms based on random walks, which have been previously published in the literature on the topic. When comparing random walk-based algorithms to other techniques, they have the benefit of being able to include a large amount of contextual information in their calculations. To offer recommendations, cooperative filtering is a kind of link prediction and recommender system that makes an attempt to locate the k-most-close nodes to a given node. This is why random walks are valuable in a range of applications, such as link prediction and recommendation systems, among others, and may be found everywhere. Various applications, such as computer vision, semi-supervised learning, network embedding, and complicated social network research, have benefited from the usage of random walks. There are many topics that academics are interested in, but some of the more popular ones include random walks on graphs, text analysis, the study of science, and knowledge discovery. There are plenty others. It is common practice to use quantum walks to increase the performance of classical algorithms by enabling them to execute at a quicker rate. It may be used for a wide range of tasks, including enhancing decision trees, solving search difficulties, and determining element distinctness, to name a few [32].

2.9   HYBRID CLASSICAL-QUANTUM ALGORITHMS

A quantum speedup produced by using the quantum walk approach does not necessarily convert into a quantum advantage in terms of the amount of computing effort required. The quadratic method has been demonstrated to produce the greatest feasible speedup while searching an unsorted database, and it outperforms the best known classical technique [33] in terms of speedup. Once it is determined that the problem has a structure that can be further investigated, it is assumed that the best classical solution would outperform random guesses when compared to those who employ the latter.

Real-world challenges are often characterised by the presence of such connections, which makes it difficult to deal with them on a practical level in practice. If you are dealing with the spin glass ground state issue [34], the most well-known classical technique is preferable to the pure quantum walk algorithm since it is less time-consuming. Using a hybrid discrete-time quantum walk-classical strategy, Montanaro [35] has shown that it outperforms both techniques independently [34,35] when used in conjunction. The following is an example of a technique for acquiring a significant quantum advantage.

Find the subroutine that is causing the problem by using the best classical method that is available to you. If the original subroutine is amenable to a quantum speedup, it is recommended that a quantum type of the subroutine be replaced for the original type. A further example may be found in the work of Montanaro [35], in which he applies this strategy to a collection of backtracking algorithms in a completely different setting. Using quantum walk search subroutines resulted in a quadratic increase in speed when compared to the usual way of searching. A benefit is that Chancellor gives case studies that explain how to develop hybrid algorithms in a continuous-time framework as an additional bonus [36,37].

Quantum computers are being used for the bottlenecks in classical algorithms, which goes in nicely with the recent diversity in computer hardware. As they approached the speed restrictions of conventional silicon-based computer processors, they have gotten increasingly sophisticated to speed up the calculation of typical tasks. Among the most common instances is graphics coprocessors. They began by rapidly refreshing the screen image and progressed to scientific computing as highly parallel devices with shared memory space. A specialised chip for communicating duties such as ethernet and Wi-Fi is often included in modern computers. In addition to several traditional CPU cores, a typical powerful computational centre now contains Graphical Processing Units (GPUs) and Field Programmable Gate Arrays (FPGAs). A quantum co-processor is a logical continuation in the same fashion, allowing for the execution of hybrid quantum-classical algorithms with well-integrated hardware.

Graph engineering, preliminary state preparation, and state transmission are all aspects of quantum walks involving dynamical control. Multi-particle quantum walks have been demonstrated to be the potential of universal quantum computation. From nuclear magnetic resonance and trapped atoms to linear optics, quantum walks have been used in a variety of physical systems. Walks in greater dimensions, in complicated networks, and with several walkers, including entangled walkers, have all been studied recently.

2.10 QUANTUM MECHANICS

Quantum computation is a multidisciplinary research area concerned with the design of quantum computers as well as quantum data processing systems, i.e., computers and data processing systems that take advantage of nature's quantum mechanical features.

The creation of new and robust computation methods enables to help enhance the processing capability for tackling specific challenges. There is a growing use of quantum computing in a variety of fields of technology like image processing and computational geometry, pattern recognition, quantum gaming, and warfare. Simulating complicated physical systems as well as mathematical issues for which there is no known traditional digital computer technique that can proficiently simulate them, quantum computers can be more helpful.

Because quantum physics is a difficult topic to understand, developing good quantum algorithms is a difficult task. Aside from that, intuition plays an important role in the creation of algorithms. An algorithm that is capable of more than merely accomplishing the job at hand is required by a quantum algorithm, as opposed to a classical algorithm, which must be capable of doing something better or being more effective in comparison to another classical algorithm. The use of quantum walks, which are the quantum mechanical equivalents of conventional random walks, to build quantum algorithms has become a prevalent method in the field of quantum computing. Quantum walks are the quantum mechanical analogues of conventional random walks, and they are used to study quantum mechanics. Recently, according to the research team, it has been shown that there is a universal quantum computing paradigm at work.

When quantum mechanics was initially proposed in the early 1900s, it was used to explain nature at the atomic level, paving the way for technological advancements such as transistors, laser-based imaging, magnetic resonance imaging (MRI), and other kinds of MRI. Physicist Richard Feynman argued in 1982 that classical logic computing was incapable of processing computations representing quantum occurrences in an amenable manner [38], and this was the first time that the relationship between quantum mechanics and information theory received widespread attention.

As it turns out, a large number of our intuitive thoughts and conceptions, as well as the non-mathematical language that we use to express them, arose prior to the research of subatomic events [39], which adds to the discussion over alternative interpretations of the results. Some well-established concepts and definitions of words are pre-quantum mechanical when seen from this viewpoint, and as a consequence, they may need revision. With regards to mathematics, quantum mechanics is a field of research that investigates the nature of phenomena by using a common mathematical language to do so. In non-mathematical terms, the fact that some events are recognised as such does not indicate that there is agreement on how to characterise them. In quantum physics literature, there are many different meanings for some sentences; it is critical to be aware of all of these interpretations before advancing. It is common for authors to ignore the necessity of clearly identifying the message they want to convey via their writing. If even the tiniest error is introduced into the non-mathematical language, it is possible that understanding quantum physics will prove to be more difficult than it should be. According to Ref. [40], learning about quantum mechanics may be performed via a number of methods and techniques. Yet, no strategy has been successful in persuading the vast majority of physicists [41]. Many techniques indicate to places where further study is needed to delve deeper into the details of quantum mechanics. Examples include putting pre-quantum mechanical notions and intuition to the test, as well as using non-mathematical terminology with more correctness and rigour in non-mathematical contexts.

There is also a broad consensus that confirming quantum mechanics prescription is virtually usually a statistical method. Individual events are prescribed by quantum mechanics in the simplest form of not casting them off. It also specifies probabilities of 1 or 0, which are used to specify specific events. Quantum mechanics, on the other hand, simply prescribes regularities among numerous events.

Quantum mechanics prescribes everything in terms of occurrences. The instantiation of single quality or multiple qualities inside any region in the spacetime is referred to as an event in this review. The importance of events in quantum mechanics has led some to speculate that events, not systems, are the underlying ontology of independent reality [38].

This method does not preclude the idea of a structure from being valuable in studying quantum mechanics. However, this raises an idea that such systems are made up of discrete occurrences instead of existing in a continuous state.

The use of quantum mechanics laws has the further benefit of allowing the unitary power to be strictly greater than that of the classical one. The search for quantum algorithms is driven by a desire to learn more about this newfound potential.

2.11   RELATIONSHIPS BETWEEN QUANTUM WALK AND QUANTUM MECHANICS

With regard to the link between quantum walk models and relativistic quantum physics, there has been a considerable lot of discussion and disagreement. Based on the work of Feynman and Hibbs, the connection between quantum walk and relativistic quantum mechanics can be traced back to the discrete formulation of the one-dimensional Dirac equation propagator [42], which was developed by Feynman and Hibbs and established the link between quantum walk and relativistic quantum mechanics. In further work, Bialynicki-Birula and Meyer [43,44] showed that there are similarities between relativistic wave equations and unitary cellular automata, as well as parallels between unitary cellular automata and quantum lattice gas automata, among other things. A link between unitary cellular automata and quantum lattice gas automata was also found by Bialynicki-Birula and Meyer [44] in their research.

Quantum walks [42] are the quantum counterparts of classical random walks, and they are constructed by using quantum physics principles such as interference and superposition to achieve their effects. They are the quantum equivalents of conventional random walks. In the next sections, we'll go into further detail regarding each of them.

2.11.1 SUPERPOSITION

If we consider quantum physics, it is described as the paradoxical ability of a quantum entity, such as an electron, to exist in several states at the same time under a variety of different circumstances, such as extreme temperatures. Depending on the scenario, one of them may indicate the lowest energy level of an atom, and the other may represent the maximum energy level of an atom. As a result of these dualistic conditions, it will be conceivable for an electron to exist in both the lowest and highest states at the same time if it is formed in a superposition as a result of these dualistic circumstances and gets imprisoned between them. A measurement that causes the superposition to be disrupted may only be determined if the object is in the lowest or highest state of the superposition, depending on the measurement. According to some speculation, the study of superposition may lead to a better understanding of the qubit, which is a fundamental unit of information in quantum computing and may be used to calculate quantum states. Traditional computing views bits as transistors that may be turned on or off, with the on and off states corresponding to the numbers 0 and 1. The numerals 0 and 1 in qubits, which are made up of electrons, may simply represent the energy levels of electrons at their lowest and greatest conceivable values respectively, whereas traditional bits must be in either the 0 or 1 state. Qubits, in contrast to conventional bits can exist in superpositions with changing probability, which can be adjusted by quantum operations while calculations are being performed, whereas traditional bits must be in either the 0 or 1 state, depending on the circumstances. Another advantage of qubits is that they are more efficient in terms of storage capacity than standard bits, which is another benefit.

2.11.2 ENTANGLEMENT

Entanglement happens once quantum entities are produced and managed in such a way that nothing of these could be clarified deprived of reference to the other such. Specific identities have been shattered. At the time you examine that entanglement may continue above great distance, this notion is quite hard to decipher. It appears as if data can move faster than the speed of light when one member of an entangled pair measures its companion [45].

2.12 QUANTUM COMPUTERS AND CIRCUITS

The laws of quantum mechanics govern the operation of a quantum computer. Its computations must be unitary transformations upon the state space.

2.12.1 QUBITS

The initial concept underneath quantum algorithms was just to begin by initialising a qubit set with applying (single or multiple) evolution operators numerous stages deprived of creating intermediary quantities, as these quantities were only for performance to be done at the conclusion of the computational procedure [46].

The first quantum algorithms grounded on quantum walks, predictably, followed the identical tactic: initialise qubits, use evolution operators, and then detect merely to compute the algorithm's conclusion. Indeed, this strategy has shown to be extremely effective in the development of a number of exceptional algorithms [47]. Furthermore, as the research has progressed, it was noted that making (partial) measurement on a quantum walk might lead to intriguing mathematical features that can be used to develop algorithms.

A classical computer works with sequences of bits called qubits. A quantum computer, on the other hand, works with qubits. Each qubit is a dual-level method with dual states across it. The tensor product of n $2d$ qubit spaces can be thought of as the state Hilbert space of n qubits. Another method to consider the states is to think of them as they are connected by n-bit strings $x_1, x_2, \ldots, x_{ni}$ with $x_i \in \{0, 1\}$. The quantum computer can be in any state of any combination of these fundamental states. In the case of a physicist, it could be cooler to consider a set of n spins half the total number of particles.

2.12.2 GATES

According to the application, a circuit consisting of gates (such as AND and NOT) that operate on one or two bits may be used to represent a calculation that may be done by a classical computer with one or two bits. In order to calculate a Boolean function from an input state, it is required to first set the computer to the appropriate input state before attempting to calculate the function. To distinguish between a quantum computer and a normal computer, one technique that has been proposed is to use a circuit made up of unitary gates that can operate with one or two qubits. This circuit can be used to distinguish between the two types of computers. Our goal is to convert the starting state into a unitary transformation that can be applied to any other starting state in the process.

Another way of putting it is that each Boolean function may be broken down into a series of gates that are selected from a universal set, which is well-known in the fields of computer science and engineering (like AND and NOT). In a typical computer, each calculation is expressed as a series of AND and NOT statements, which allows for a more clear explanation. Since the discovery that any unitary computation on qubits can be broken down into a series of quantum gates that can only operate on single or dual qubits, it has been celebrated as one of the most important breakthroughs in current quantum computing theory and practice.

In order to create a universal quantum gate collection, unitary matrices with a single qubit, such as the single matrix and the CNOT matrix, are utilised. Both of these matrices have a single qubit and are examples of unitary matrices that may be employed.

The first stage in building a quantum computer is to design a device that is capable of performing single-qubit unitaries as well as a two-body transformation compatible with the CNOT algorithm. The second part of the procedure entails the creation of a device that is capable of performing two-body transformations that are compatible with the CNOT algorithm, as described above. According to the researchers, this initiative will investigate whether or not a quantum computer, including its unitaries, can execute computing tasks that are comparable to those done by conventional computers. The reversibility of unitary transformations implies that every quantum calculation is also reversible; in fact, the reversibility of unitary transformations implies that it is feasible for a session of quantum computations to include no classical circuits at all.

Bennett [7] has proved that this is not a problem: any classical circuit can be made reversible by replacing its gates with reversible three-bit gates and increasing the amount of scratch space available in the circuit. Having a quantum computer capable of completing these reversible gates in a unidirectional method is also advantageous. In light of the fact that a quantum computer is capable of doing any classical computation with ease, it follows that quantum computers must be at least as powerful as their classical counterparts in order to be useful in the real world. Quantum computers, on the other hand, are not yet as powerful as their classical counterparts.

2.13 DRONE-BASED QUANTUM COMPUTING

Quasi-quantum satellite networks are being investigated by researchers in China and Europe, with the hope of making it possible for satellites to communicate with the earth across very great distances in the future. Despite this, there are a number of downsides to using quantum satellites that must be taken into mind. However, despite the fact that low-orbit satellites are capable of interacting with particular soil locations within a certain amount of time, the excessive cost of space launch makes the building of a quantum satellite system prohibitively expensive. In the future, they hope to be able to develop the world's first global quantum internet, which would be based on quantum particle transmission, thanks to their ability to generate secret codes for encrypting data using particles that interact in an ultra-secure manner. According to the hypothesis, a quantum web might also enable distant quantum computers to interact or conduct research at the boundaries of quantum physics via the utilisation of a network of quantum computers. There has already been progress in the delivery of photons across China through a quantum satellite and the usage of fibre-optical cable quantum networks. A number of variables, including mobility, relative simplicity of operation, and cheap cost, suggest that drones have the potential to be employed as an alternative technology.

Researchers at Nanjing University in China have developed a "quantum drone" that will serve as an aerial component in a quantum network, highlighting the current surge in drone technology. "It was employed as a node and produced the first quantum drone identical to the quantum satellites used," said a quantum physicist of the University of Rostock in Germany, who was not engaged in this research. "Drones can be deployed at any location and time for a mobile quantum link," said a member of the study team from the University of Nanjing. Drones can easily be adjusted to avoid fog and pollution. They may also be less expensive and further flexible when compared to satellite quantum systems, necessitating a focus on drone-based quantum technology study.

2.14 QUANTUM SATELLITES AIMED AT DRONE-BASED NETWORK AS WELL AS COMMUNICATIONS

The goal of quantum satellites is to shape the future of wireless communication and to take steps to improve internet security. In August 2016, China [48] successfully deployed the first quantum satellite into space. The quantum satellite was launched into orbit for testing, and the results showed that the technology is reliable and unhackable. Quantum entanglement, in which subatomic particles get intrinsically tied or "entangled", was utilised to enable satellites to communicate with the earth. Because they are on opposing sides of the cosmos, modifying one will prevent the other from modifying. This proves that although entangled particles are tried, accessing them will be difficult.

In 2017, the advancement was furthered by the demonstration of entanglement-based quantum key distribution, which reduced the focus on the error recognition rates while enabling secure communications [49]. Increased light-gathering accuracy and updated filtering systems and other optical elements were incorporated into the improvements, resulting in a low error rate for quantum key distribution.

2.15 CONCLUSIONS

Quantum walks are common models for physical processes, particularly transport features. Quantum algorithms can also benefit from them. It is critical to comprehend the various needs for various quantum walk applications. Thankfully, quantum walks and quantum mechanics are now a well-established field of quantum computing study, with a plethora of intriguing open challenges for physicists, computer scientists, mathematicians, and engineers to solve. This chapter, which is intended to be an aid to the areas of quantum walks and mechanics, will better benefit the technical world if it motivates quantum scientists as well as quantum engineers to pursue additional research in this field.

The fundamentals of quantum computation have been established, and everything needed for its future development is still being researched. Quantum algorithms, logic gate operations, error checking, decoherence dynamics and controls, atomic-scale technologies, and relevant applications can further be explored in depth. Few concepts based on the quantum walk and quantum mechanics have been presented, with the goal of providing a simple yet broad solution to the situation. In the future, better models are to be projected to make an impact by researchers in quantum physics. Although the future cannot be predicted, it is bright in the field of quantum physics.

REFERENCES

[1] N. Gisin, G. G. Ribordy, W. Tittel, and H. Zbinden. Quantum cryptography. *Rev. Mod. Phys.*, 74(1):145–195, 2002.

[2] P. W. Shor. Polynomial-time algorithms for prime factorization and discrete logarithms on a quantum computer. *SIAM J. Comp.*, 26(5):1484–1509, 1997.

[3] Y. Aharonov, L. Davidovich, and N. Zagury. Quantum random walks. *Phys. Rev. A*, 48(2):1687–1690, 1993.

[4] G.Piccinini.Computationinphysicalsystems.InE.N.Zalta(Ed.),*TheStanfordEncyclopedia of Philosophy*, Summer 2017 edition, Stanford University Press. 2017. Available at http:// plato.stanford.edu/archives/sum2017/entries/computation-physicalsystems/

[5] C. Horsman, S. Stepney, R. C. Wagner, and V. Kendon. When does a physical system compute? *Proc. Roy. Soc. A*, 470(2169):20140182, 2014. doi:10.1098/rspa.2014.0182

[6] R. Blume-Kohout, C. M. Caves, and I. H. Deutsch. Climbing mount scalable: physical resource requirements for a scalable quantum computer. *Found. Phys.*, 32(11):1641–1670, 2002. doi:10.1023/A:1021471621587

[7] C. H. Bennett. Logical reversibility of computation. *IBM J. Res. Develop.*, 17:5225, 1973.

[8] N. Shenvi, J. Kempe, and R.B. Whaley. A quantum random walk search algorithm. *Phys. Rev. A*, 67(5):052307, 2003.

[9] A. M. Childs, R. Cleve, E. Deotto, E. Farhi, S. Gutmann and D. A. Spielman. Exponential algorithmic speedup by a quantum walk. In *Proceedings of the Thirty-Fifth Annual ACM Symposium on Theory of Computing*, pp. 59–68, 2003.

[10] A. Ambainis. Quantum walks and their algorithmic applications, *Int. J. Quantum Inf.*, 1:507–518, 2003.

[11] J. Kempe. Quantum random walks: an introductory overview, *Contemp. Phys.*, 44:307–327, 2003.

[12] L. Page, S. Brin, R. Motwani, and T. Winograd. *The PageRank Citation Ranking: Bringing Order to the Web*. Stanford InfoLab, 1999.

[13] D. Fogaras, B. Rácz, K. Csalogány, and T. Sarlós. Towards scaling fully personalized PageRank: algorithms, lower bounds, and experiments, *Internet Math.*, 2(3):333–358, 2005.

[14] T. H. Haveliwala, Topic-sensitive PageRank: a context-sensitive ranking algorithm for web search, *IEEE Trans. Knowl. Data Eng.*, 15(4):784–796, 2003.

[15] J.-Y. Pan, H.-J. Yang, C. Faloutsos, and P. Duygulu, Automatic multimedia cross-modal correlation discovery. In *Proceedings of the Tenth ACM SIGKDD International Conference on Knowledge Discovery and Data Mining*, pp. 653–658, 2004.

[16] J. Shen, Y. Du, W. Wang, and X. Li. Lazy random walks for superpixel segmentation, *IEEE Trans. Image Process.*, 23(4):1451–1462, 2014.

[17] A. Plavnick. The fundamental theorem of Markov chains, VIGRE REU at UChicago, 2008.

[18] J. Kempe. Quantum random walks - an introductory overview, arXiv:quant-ph/0303081, 2003.

[19] A. Sinclair. *Algorithms for Random Generation and Counting, A Markov Chain Approach*. Birkhauser Press, 1993.

[20] M. A. Broome, A. Fedrizzi, B. P. Lanyon, I. Kassal, A. Aspuru-Guzik, and A. G. White. Discrete single-photon quantum walks with tunable decoherence. *Phys. Rev. Lett.*, 104:153602, 2010. doi:10.1103/PhysRevLett.104.153602

[21] W. Dür, R. Raussendorf, V.M. Kendon, and H.-J. Briegel. Quantum random walks in optical lattices. *Phys. Rev. A*, 66:052319, 2002.

[22] N. Konno. Limit theorems and absorption problems for quantum random walks in one dimension. *Quantum Inf. Comput.*, 2:578–595, 2002

[23] R. P. Feynman and A. R. Hibbs. *Quantum Mechanics and Path Integrals*. International Series in Pure and Applied Physics. McGraw-Hill, New York, 1965.

[24] D. Meyer. From quantum cellular automata to quantum lattice gases. *Journal of Statistical Physics.*, Springer link https://doi.org/10.1007/BF02199356, 85:551–574, 1996.

[25] J. Watrous. Quantum simulations of classical random walks and undirected graph connectivity. *J. Comput. Syst. Sci.*, 62(2):376–391, 2001.

[26] A. Ambainis, E. Bach, A. Nayak, A. Vishwanath, and J. Watrous. One-dimensional quantum walks. In *Proc. 33th STOC*, ACM, New York, pp. 60–69, 2001.

[27] E. Farhi and S. Gutmann. Quantum Computation and Decision Trees, arXiv:quant-ph/9706062v2, 1998.

[28] N. Shenvi, J. Kempe, and K. Whaley. A Quantum Random Walk Search Algorithm, arXiv:quantph/0210064v1, 2008.

[29] K. Wiesner. Quantum cellular automata. In R. A. Meyers (Ed.), *Encyclopedia of Complexity and System Science*, Springer, pp. 7154–7164, 2009.

[30] S. Aaronson and A. Arkhipov. The computational complexity of linear optics. *Theory Comput.*, 9(4):143–252, 2013. doi:10.4086/toc.2013.v009a004

[31] A. B. Dodds, V. Kendon, C. S. Adams, and N. Chancellor. Practical designs for permutation-symmetric problem Hamiltonians on hypercubes. *Phys. Rev. A*, 100:032320, 2019. doi:10.1103/PhysRevA.100.032320

[32] S. E. Venegas-Andraca. Quantum walks for computer scientists. *Synth. Lect. Quantum Comput.*, 1(1):1–19, 2008.

[33] C. Bennett, E. Bernstein, G. Brassard, and U. Vazirani. Strengths and weaknesses of quantum computing. *SIAM J. Comput.*, 6(5):1510–1523, 1997. doi:10.1137/S0097539796300933

[34] A. Callison, N. Chancellor, F. Mintert, and V. Kendon. Finding spin glass ground states using quantum walks. *New J. Phys.*, 21:123022, 2019. doi:10.1088/1367–2630/ab5ca2

[35] A. Montanaro. Quantum speedup of branch-and-bound algorithms, 2019. Available at http://arxiv.org/abs/1906.10375. ArXiv:1906.10375.

[36] N. Chancellor. Modernizing Quantum Annealing II: Genetic algorithms with the Inference Primitive Formalism, 2017. Available at https://arxiv.org/abs/1609.05875. ArXiv:1609.05875.

[37] N. Chancellor. Modernizing quantum annealing using local searches. *New J. Phys.*, 19(2):023024, 2017. doi:10.1088/1367–2630/aa59c4.

[38] T. Maudlin. Ontological clarity via canonical presentation: electromagnetism and the Aharonov-Bohm effect. *Entropy*, 20:465, 2018.

[39] H. H. Grelland. The Sapir-Whorf hypothesis and the meaning of quantum mechanics, In: G.A. Adenier, A. Khrennikov, T.M. Nieuwenhuizen (Eds.), *3th Conference on Quantum Theory: Reconsideration of Foundations, Växjö, Sweden, 6–11 Jun 2005*. American Institute of Physics, New York, pp. 325–329, 2006.

[40] A. Cabello. Interpretations of quantum theory: a map of madness, In O. Lombardi, S. Fortin, F. Holik, and C. López (Eds.), *What Is Quantum Information?* Cambridge University Press, Cambridge, pp. 138–144, 2017.

[41] M. Schlosshauer, J. Kofler, A. Zeilinger. A snapshot of foundational attitudes toward quantum mechanics. *Stud. Hist. Philos. Mod. Phys.*, 44:222–230, 2013.

[42] R. P. Feynman and A. R. Hibbs. *Quantum Mechanics and Path Integrals*. McGraw-Hill, New York, 1965.

[43] I. Bialynicki-Birula. Weyl, Dirac, and Maxwell equations on a lattice as unitary cellular automata, *Phys. Rev. D*, 49:6920, 1994.

[44] D. A. Meyer. From quantum cellular automata to quantum lattice gases, *J. Stat. Phys.*, 85:551, 1996.

[45] M. Born. *The Born-Einstein Letters*. Walker, London, 1971.

[46] D. Bouwmeester, A. Ekert, and A. Zeilinger (Eds.). *The Physics of Quantum Information*. Springer, 2001.

[47] A. Ambainis. Quantum search algorithms. *SIGACT News*, 35:22–35, 2004

[48] Q. Zhang, F. Xu, L. Li, N.-L. Liu, and J.-W. Pan. Quantum information research in China, *Quantum Sci. Technol.*, 4(4):040503, 2019. doi: 10.1088/2058–9565/AB4BEA

[49] S. Liao, J. Lin, J. Ren, W. Liu, J. Qiang, J. Yin, Y. Li, Q. Shen, L. Zhang, X. Liang, and H. Yong. Space-to-ground quantum key distribution using a small-sized payload on Tiangong-2 space lab. *Chin. Phys. Lett.*, 34(9):90302, 2017.

3 A Keen Study on Quantum Information Systems

Rupam Bhagawati and Thiruselvan Subramanian
Presidency University

Arish Pitchai
Entropik Technologies Pvt. Ltd

CONTENTS

3.1 INTRODUCTION

The best-known expert of quantum mechanics, Albert Einstein, went to his grave unreconciled with the hypothesis he developed. Generations of physicists since have wrestled with quantum mechanics with an end goal to make its predictions more satisfactory. Quantum mechanics is a mathematical design or set of rules for the development of actual hypotheses. For example, there is an actual speculation known as quantum electrodynamics which portrays with awesome accuracy the cooperation of particles and light. Quantum electrodynamics is created inside the construction of quantum mechanics, yet it contains explicit not totally settled by quantum mechanics. The relationship of quantum mechanics to explicit actual hypotheses like quantum

DOI: 10.1201/9781003250357-3

electrodynamics is genuinely like the relationship of a Personel Computer's functioning framework to explicit applications programming—the functioning framework sets explicit essential boundaries and methods of action but leaves open how explicit undertakings are achieved by the applications. The guidelines of quantum mechanics are straightforward yet even specialists imagine that they are irrational; furthermore, the earliest forerunners of quantum calculation and quantum information may be found in the long-standing craving of physicists to all the more promptly get quantum mechanics.

Quantum registering is viewed as the fate of processing and the distinct advantage for a ton of fields relying upon it, i.e., from cryptography to medication, the entire way to applied sciences, and software engineering. Quantum processing is a review distinct that spotlights creating computers in view of the properties of "quantum hypothesis" and "quantum physical science". Quantum hypothesis portrays the lead and nature of energy and matter on the "nuclear" and "subatomic" levels, in any case, called the quantum levels. A quantum framework is a computational contraption that utilizes the eccentricities of quantum mechanics with its quantum properties to plan and deal with data. A tremendous scope quantum PC is prepared for performing tasks in very high speed which is much more than a classical method. Improvement of such computers is viewed as an enormous leap for humankind in processing capacities with gigantic execution gains for regenerations or savage power for example [1].

3.1.1 TERMINOLOGY IN QUANTUM COMPUTATION

Qubits address particles of an atom and a qubit is seen as the "central unit of information" for quantum frameworks. In conventional frameworks, information is encoded in bits that have either 1 or 0 as a value. Thusly, a piece should be in one of the either states, 1 or 0. Be that as it may, this isn't valid for qubits as qubits are not limited to being in one state and can exist in superposition. As such, a qubit can exist in 0, 1, or an immediate blend of the two states [2].

As frameworks isolate information into bits, a quantum framework wouldn't use a conventional switch circuit. In light of everything, a quantum real structure relies upon the way that qubits can exist in 0 and 1 simultaneously yet this could struggle with how we could decipher regular actual science. The solid computational power of a quantum PC should be achieved if qubits go through "entanglement", which is the most widely recognized approach to trapping qubits into social events (quantum registers) making an amazingly extreme information-processing gear [3,4].

> **Quantum Search:** Quantum calculations offer "polynomial speedup" over the most popular old-style calculation for different issues particularly the "quantum data set search" that can have handled use "Grover's Algorithm" that purposes "quadratic partner" less inquiries to look in an information base contrasted with a traditional computer consequently ending up "ideal" rather than rigorously a proof of idea [5,6].

Quantum Simulation: Nanotechnology and science researchers are profoundly expecting the utilization of quantum recreations to re-enact complex trials that are incomprehensible on an old-style computer [7].

Quantum PC can apply computational strides to its register of qubits. Two models exist for this: the quantum Turing machine and the quantum circuit model [8].

Entanglement Frontier: Quantum information science investigates the past of outstandingly complex quantum expresses, the "entanglement wild". Traditional frameworks can't reproduce significantly caught quantum frameworks successfully, and we want to hurry the day when generally around controlled quantum frameworks can perform assignments outperforming what should be conceivable in the old-style world. One strategy for achieving such "quantum incomparability" is to run an estimation on a quantum framework which settles an issue with a super-polynomial speedup similar to old-style computers, but there may be substitute ways that can be accomplished sooner, for example, mimicking fascinating quantum conditions of unequivocally related matter [9]. To work a colossal scale quantum PC reliably, we should overcome the weakening impacts of cognizance, which might have done utilizing "standard" quantum gear defended by quantum botch changing codes, or by exploiting the non-abelian quantum insights of annoys recognized in strong state frameworks, or by joining the two techniques. Basically, by testing the entanglement backwoods will we understand whether nature gives outrageous assets quite far past what the old-style world would allow.

Superposition Synthesis: The quantum search calculation can be looked at as a technique for orchestrating a particular kind of superposition-one whose adequacy is accumulated in a solitary premise state. This premise state is portrayed by a twofold capacity $f(\overline{x})$ that is non-zero in this ideal premise state and zero everywhere else [10].

3.2 QUANTUM INFORMATION PROCESSING

The essential consequences of traditional information speculation are Shannon's quiet channel coding theory and Shannon's loud channel coding theory [11]. The silent channel coding theory evaluates the quantity of pieces that are relied upon to store information being released by a wellspring of information, while the loud channel coding speculation measures how much information can be constantly sent through a boisterous interchanges channel. What do we mean by an information source? Describing this thought is a focal issue of traditional and quantum information theory, one we'll reconsider a few times [12].

Shannon's noiseless coding theory gives a certifiable model where the objectives of information speculation recorded before are completely met where two static assets are recognized as the piece of information source. A two-stage dynamic cycle is perceived as packing an information source, and a while later de-pressurizing to recover the information source. Finally, a quantitative reason for concluding the assets consumed by an ideal information pressure plot is found.

Shannon's second huge outcome, the uproarious channel coding theory, evaluates how much information can be reliably sent through a boisterous channel. In particular, assume we wish to move the information being conveyed by some information source to another region through a boisterous channel. That region may be at another point in space, or at another event—the last choice is the issue of putting away information within the sight of commotion. The idea on the two occasions is to encode the information being made utilizing botch changing codes so that any commotion introduced by the channel can be updated at the contrary completion of the channel [12,13].

Quantum data processors take advantage of the quantum highlights of superposition and entrapment for implementation crazy in old-style contraptions, offering the potential for colossal redesigns in the correspondence and handling of data [14]. Preliminary insistence of huge extension quantum data processors remains a truly lengthy vision, as the fundamental for all intents and purposes unadulterated quantum lead is seen especially in fascinating stuff, for instance, individual laser-cooled particles and detached photons [14].

Claude Shannon found how to assess information—an outcome so key and moderate that looking back it is amazing it had not been arranged previously. The corresponding piece, or twofold digit, transformed into the central unit of data, giving an estimation to taking a gander at types of information and propelling how much assets are expected to faithfully pass on a given proportion of data, indeed within the sight of commotion [15].

Shannon's initial task unknowingly introduced the data age. Pieces are in natural form ungainly tubes during the 1940s to the high-level semiconductor of under 5–10 cm in size. Tragically, the days (or significant length) of Moore's law are numbered. As pieces constantly wither, they will at last push towards the size of individual particles—persistently 2020 expecting the current start to hold impact. In any case, in any event, expecting we could some way or another show up at the base, where quantum conditions of each molecule in a central processor have old-style bits, there would be no more space for extra gains without parting the molecule [15,16].

As per Ref. [17], quantum features of an assortment of particle assessed bits, without a doubt we can track down more space. Given as

$$|\psi> = \alpha|0> + \beta|1\rangle$$

α and β are complex amplitudes of qubit states $|0\rangle$ and $|1\rangle$. What's more, the superposition is settled into either $|0>$ or $|1>$ positive state upon estimation with individual probabilities $|\alpha|^2$ and $|\beta|^2 = 1 - |\alpha|^2$. In one sense, this is only a solitary snippet of information. The reliable amplitudes α and β pass on an endless proportion of information, similar to basic information carriers, for example, the predictable voltage set aside on capacitors. In any case, straightforward frameworks are known to encounter the evil impacts of the joined create of upheaval, rather than computerized standards like semiconductor reasoning that snare to their high or low levels through predictable assessment and info. Quantum pieces are equivalently weak against straightforward commotion, be that as it may, they can give essentially more like a compromise—quantum entanglement (Figure 3.1).

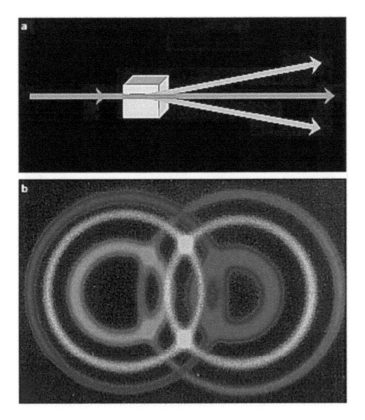

FIGURE 3.1 Optical parametric down-transformation

In a customary basic architecture, N capacitors are required to store N endless voltages. Notwithstanding, with these, the broadest is not entirely settled by 2^N independent amplitudes. A bunch of qubits in this manner has the potential for taking care of significantly more information than a comparative arrangement of customary data transporters [18,19].

In specific calculations, notwithstanding, appropriate quantum reasoning entrances make the amplitudes interfere with the objective that two or three amplitudes make due eventually [20]. Following an assessment, the result can depend upon an overall property of each 2^N information [21,22].

Representing the method by Peter Shor's figuring estimation, where a quantum PC would have the choice to factor colossal numbers significantly speedier than any realized old-style framework's computation. Speedy quantum number consideration has huge ideas in cryptanalysis, as various renowned encryption calculations, for instance, Rivest, Shamir, Adleman algorithm relies upon the weakness to factor colossal numbers [23].

As per Refs. [24,25] basic improvement was quantum botch modification, Quantum goof revision perceives and corrects slight quantum entryway botches or weak associations with the environment through abundance encoding of qubits.

3.3 QUANTUM INFORMATION SYSTEMS

The hypothesis of quantum data, that is to say, of the transmission, stockpiling, and handling of data utilizing quantum-mechanical frameworks, is presently very much evolved. It has been built to a great extent by the speculation to this on the text of components of conventional data hypothesis, which have been intended to clarify the transmission, stockpiling, and handling data utilizing traditional means [26,27]. The numerical portrayal of quantum frameworks varies generally from that of traditional states. Not at all like in the traditional case, the greater part of the data put away in a non-exclusive quantum-mechanical framework is put away as connections between sub-systems. Not exclusively is most quantum data put away as such connections; however, these can be uncommonly solid relationships, completely entrapped quantum states are the outrageous cases. For instance, aforesaid [26,27] expresses the diminished conditions of single qubits which are completely endless, though the condition of the qubit pair is completely related, that is to say, information on the condition of one qubit completed through a quantum estimation is equivalent to information on the other, as in the circumstance considered by Einstein, Podolsky, and Rosen as outlined by Bohm [28]. The best distinction in intricacy among traditional and quantum states emerges when the trap is available among parts of composite quantum frameworks. Entanglement between even two subsystems gives a clever sort of information-processing asset supplementing what can be given by traditional pieces or untangled quantum bits. This distinction is significantly more articulated on account of bigger various qubit states, which can be dispersed among various separate gatherings possibly taking part in circulated information processing. In this part, amounts describing both static and dynamical properties of quantum information are thought of [29,30].

3.3.1 QUANTUM ENTROPY

The standard proportion of the information held inside a quantum framework portrayed by the statistical operator ρ is the von Neumann entropy.

$$S(\rho) = -tr(\rho \log_2 \rho)$$

$$= -\sum_i \lambda_i \log_2 \lambda_i$$

$S(\rho)$ is non-negative and achieves maximum value for mix state. λ_i are members of the sets of eigenvalue ρ.

The von Neumann entropy assumes a part in the quantum information hypothesis undifferentiated from that played by the Shannon entropy in the conventional information hypothesis [31].

3.3.2 QUANTUM RELATIVE AND CONDITIONAL ENTROPIES

The quantum conditional entropy, comparably to the relating traditional amount, is given by

$$S(A|B) = S(A,B) - S(B)$$

$$= S(\rho_{AB} - \rho_B)$$

However, in contrast to the traditional contingent entropy, the quantum contingent entropy can become negative, demonstrating that it is feasible for quantum systems to be more sure in the joint state of two-part systems than in the states of its singular parts, as again should be visible on account of the singlet state $|\psi^-\rangle$.

The quantum relative entropy between the two states, ρ and σ, of a quantum framework is characterized as

$$S(\rho \| \sigma) = tr\left(\rho\left(\log_2 \rho - \log_2 \sigma\right)\right)$$

This quantity obeys Klein's inequality [32].

3.4 TRENDS IN QUANTUM INFORMATION SYSTEMS

3.4.1 QUANTUM CRYPTOGRAPHY

Quantum mechanics can possibly assume a significant part coming down the line for cryptology. From one perspective, it could push to the edge of total collapse a large portion of the latest things in contemporary cryptography. Then again, it offers an option for the assurance of protection whose security can't be matched by traditional means [33].

3.4.2 QUANTUM RETRODICTION

Quantum retrodiction is a hypothesis as well as standard procedure for impelling quantum states that show wide frameworks fit retrodictive issues, yet the presence of a few new creative highlights leads to few comments on more philosophical issues related to the electrical shock as well as wave work "breakdown" towards evaluation association [34,35].

3.4.3 QUANTUM KEY DISTRIBUTION

It is a critical standard of the current trend that the encryption show ought not to be founded on lack of clarity. Kerkhoff's standard communicates that a foe established all nuances of the encryption instrument and about the possible messages, and all that isn't insinuated by the enemy is separate as a secret key [36].

To depict those registers in a quantum-mechanical world, we introduce them as quantum frameworks as states contained in an even set. Structure an explanation, holds a quantum express that might be connected with the key. The situation contemplates then to two old-style registers, embedded into the quantum language, and one totally quantum register in Eve's hold. As a last resort, the three get-togethers can be portrayed from Eve's viewpoint by

$$\rho ABE_{lccq} = \sum_K \sum_{K^!} p(K,K')|K\rangle\langle K |_A \otimes | K^{\cdot}\rangle < K' |_B \otimes \rho_E^{(K,K')}$$

where $p(K, K')$ is a joint probability distribution for the classical registers of Alice and Bob [37–39].

3.4.3.1   Applications of Quantum Information Systems

Human-scale actual characteristic addresses the new, complex or straightforward, and minute regulations. In the beyond 20 years, dealt with comprehension of these tiny regulations have recommended that normal huge scope frameworks—those utilized in trend-setting innovation from semiconductors to mechanical sensors to radios—don't take advantage of what is allowed by quantum mechanics. In particular, looking at these frameworks from an information innovation viewpoint, we can say that information as addressed in quantum mechanics is more luxurious than that allowed in the traditional setting. In looking for utilizations of this part of the ordinary world, we centre around completing new strategies for metrology, secure correspondence, reproduction of actual frameworks, and even information processing [40–44].

Hypothetical research includes the following:

- Exploring the show and cut-off points of quantum gadgets, including strategies for interfacing dissimilar subsystems like particles and superconducting quantum pieces to take advantage of the properties of each;
- Creating applications for quantum advances in estimation science, correspondence, and computation, for example, approaches for nano-scale appealing detection or critical distance quantum key dispersion;
- Considering novel jobs for scattering, not simply in seeing how it limits knowledge in quantum frameworks, yet furthermore for cooling frameworks and in any event, changing blunders in a quantum calculation.

With the outstanding advancement in registering, processing quantum frameworks are evidently accepting to handling unsolved issues, which are not for classical PCs. This movement sets out a vast expanse of entryways, across basically all aspects of current life. For sure, Google has actually stood apart as really newsworthy reporting the accomplishment of uniqueness.

3.4.4   Artificial Intelligence & Machine Learning

Artificial intelligence is a conspicuous region at the present time, due to advancements in all aspects. A part of its wide applications we can see consistently in all fields like handwriting acknowledgement. Be that as it may, as the amount of utilizations extended, it transforms into a troublesome task for conventional PCs, to arrange the precision and speed. Also, that is where quantum registering can help in handling complex issues in particularly less time, which would have required customary frameworks for millennia.

3.4.5   Cybersecurity & Cryptography

The online security space as of now has been very helpless because of the expanding number of digital assaults happening across the globe, every day. In spite of the fact that organizations are laying out important security structures, the interaction becomes overwhelming and unreasonable for traditional advanced machines. In addition, as such, network safety has kept on being a crucial concern all around the planet.

3.5 CONCLUSION

The field of digitization of each data brings information processing into its realm. The field of information processing grows at a rapid speed and achieves a milestone. Quantum is a field for this process which bends towards an endless path. Many aspects of processing come out as an output for the different tasks that are included in a customized manner and already implemented by customary information processors.

REFERENCES

1. N. Anisoara and A. S. Varde, Association for Computing Machinery. Special Interest Group on Hypertext, H. and Web., Association for Computing Machinery, Special Interest Group on Information Retrieval, Association for Computing Machinery. Special Interest Group on Knowledge Discovery & Data Mining., Association for Computing Machinery., & ACM International Conference on Information and Knowledge Management, *Proceedings of the 3rd Workshop on Ph. D. Students in Information and Knowledge.* ACM, Toronto, ON (2010).
2. Association for Computing Machinery. Special Interest Group on Information Retrieval, *SIGIR '08 : The 31st Annual International ACM SIGIR Conference on Research and Development in Information Retrieval, July 20–24, 2008.* Association for Computing Machinery, Singapore (2008).
3. Y. Xiao, C. Guo, F. Meng, N. Jing, and M. H. Yung, Incompatibility of observables as state-independent bound of uncertainty relations, *Phys. Rev. A* 100(3), 032118 (2019). https://doi.org/10.1103/PhysRevA.100.032118
4. M. D. Dunlop, The effect of accessing nonmatching documents on relevance feedback, *ACM Trans. Inf. Syst.* 15(2), 137–153 (1997).
5. G. Zuccon and L. Azzopardi, Using the quantum probability ranking principle to rank interdependent documents. In *European Conference on Information Retrieval*, pp. 357–369. Springer, Berlin, Heidelberg (2010).
6. G. Zuccon, L. A. Azzopardi, and K. van Rijsbergen, The quantum probability ranking principle for information retrieval, *Lect. Notes Comput. Sci.* 5766, 232–240 (2009). https://doi.org/10.1007/978-3-642-04417-5_21
7. C. J. van Rijsbergen, *Information Retrieval.* Butterworths Journal of Documentation, Harvard (1975).
8. M. E. Maron and J. L. Kuhns, On relevance, probabilistic indexing and information retrieval, *J. ACM* 7(3), 216–244 (1960).
9. S. E. Robertson, *The Probability Ranking Principle in IR*, pp. 281–286. Morgan Kaufmann Publishers Inc., San Francisco, CA (1997).
10. M. Eisenberg and C. Barry, Order effects: A study of the possible influence of presentation order on user judgments of document relevance. *J. Am. Soc. Inf. Sci.* 39(5), 293–300, 1988.
11. R. P. Feynman, The concept of probability in quantum mechanics. In *Proceedings of the Second Berkeley Symposium on Mathematical Statistics and Probability*, pp. 533–541. University of California Press (1951).

12. C. X. Zhai, W. W. Cohen, and J. Lafferty, Beyond independent relevance: Methods and evaluation metrics for subtopic retrieval. In *SIGIR Forum (ACM Special Interest Group on Information Retrieval)*, pp. 10–17, ACM, New York (2003).

13. W. Heisenberg, About the descriptive content of quantum theoretical kinematics and mechanics, *Z. Phys.* **43**, 172 (1927).

14. E. H. Kennard, To quantum mechanics of simple types of movement, *Z. Phys.* **44**, 326 (1927).

15. H. Weyl, Gruppentheorie und Quantenmechanik (Hirzel, Leipzig, 1928) [English translation: H. P. Robertson, The Theory of Groups and Quantum Mechanics (Dover, New York, 1931)].

16. H. P. Robertson, The uncertainty principle, *Phys. Rev.* **34**, 163 (1929).

17. E. Schrödinger, On Heisenberg's uncertainty principle, Meeting reports of the Prussian Academy of Sciences, physical-mathematical class, 14, 296 (1930).

18. H. F. Hofmann and S. Takeuchi, Violation of local uncertainty relations as a signature of entanglement, *Phys. Rev. A* **68**, 032103 (2003).

19. O. Gühne, Characterizing entanglement via uncertainty relations, *Phys. Rev. Lett.* **92**, 117903 (2004).

20. J. Oppenheim and S. Wehner, The uncertainty principle determines the nonlocality of quantum mechanics, *Science* **330**, 1072 (2010).

21. M. D. Reid, Demonstration of the Einstein-Podolsky-Rosen paradox using nondegenerate parametric amplification, *Phys. Rev. A* **40**, 913 (1989).

22. E. G. Cavalcanti, S. J. Jones, H. M. Wiseman, and M. D. Reid, Experimental criteria for steering and the Einstein-Podolsky-Rosen paradox, *Phys. Rev. A* **80**, 032112 (2009).

23. A. Rutkowski, A. Buraczewski, P. Horodecki, and M. Stobińska, Quantum steering inequality with tolerance for measurement-setting errors: experimentally feasible signature of unbounded violation, *Phys. Rev. Lett.* **118**, 020402 (2017).

24. Z.-A. Jia, Y.-C. Wu, and G.-C. Guo, Characterizing nonlocal correlations via universal uncertainty relations, *Phys. Rev. A* **96**, 032122 (2017).

25. Y. Xiao, Y. Xiang, Q. He, and B. C. Sanders, Quasi-fine-grained uncertainty relations, arXiv:1807.07829.

26. D. Deutsch, Uncertainty in quantum measurements, *Phys. Rev. Lett.* **50**, 631 (1983).

27. M. H. Partovi, Entropic formulation of uncertainty for quantum measurements, *Phys. Rev. Lett.* **50**, 1883 (1983).

28. K. Kraus, Complementary observables and uncertainty relations, *Phys. Rev. D* **35**, 3070 (1987).

29. H. Maassen and J. B. M. Uffink, Generalized entropic uncertainty relations, *Phys. Rev. Lett.* **60**, 1103 (1988).

30. I. D. Ivanovic, An inequality for the sum of entropies of unbiased quantum measurements, *J. Phys. A* **25**, L363 (1992).

31. J. Sánchez, Entropic uncertainty and certainty relations for complementary observables, *Phys. Lett. A* **173**, 233 (1993).

32. M. A. Ballester and S. Wehner, Entropic uncertainty relations and locking: Tight bounds for mutually unbiased bases, *Phys. Rev. A* **75**, 022319 (2007).

33. S. Wu, S. Yu, and K. Mølmer, Entropic uncertainty relation for mutually unbiased bases, *Phys. Rev. A* **79**, 022104 (2009).

34. M. Berta, M. Christandl, R. Colbeck, J. M. Renes, and R. Renner, The uncertainty principle in the presence of quantum memory, *Nat. Phys.* **6**, 659 (2010).

35. C.-F. Li, J.-S. Xu, X.-Y. Xu, K. Li, and G.-C. Guo, Experimental investigation of the entanglement assisted entropic uncertainty principle, *Nat. Phys.* **7**, 752 (2011).

36. R. Prevedel, D. R. Hamel, R. Colbeck, K. Fisher, and K. J. Resch, Experimental investigation of the uncertainty principle in the presence of quantum memory and its application to witnessing entanglement, *Nat. Phys.* **7**, 757 (2011).

37. Y. Huang, Entropic uncertainty relations in multidimensional position and momentum spaces, *Phys. Rev. A* **83**, 052124 (2011).
38. M. Tomamichel and R. Renner, Uncertainty relation for smooth entropies, *Phys. Rev. Lett.* **106**, 110506 (2011).
39. P. J. Coles, R. Colbeck, L. Yu, and M. Zwolak, Uncertainty relations from simple entropic properties, *Phys. Rev. Lett.* **108**, 210405 (2012).
40. P. J. Coles and M. Piani, Improved entropic uncertainty relations and information exclusion relations, *Phys. Rev. A* **89**, 022112 (2014).
41. J. Kaniewski, M. Tomamichel, and S. Wehner, Entropic uncertainty from effective anticommutators, *Phys. Rev. A* **90**, 012332 (2014).
42. Y. Xiao, N. Jing, S.-M. Fei, T. Li, X. Li-Jost, T. Ma, and Z.-X. Wang, Strong entropic uncertainty relations for multiple measurements, *Phys. Rev. A* **93**, 042125 (2016).
43. Y. Xiao, N. Jing, S.-M. Fei, and X. Li-Jost, Improved uncertainty relation in the presence of quantum memory, *J. Phys. A* **49**, 49LT01 (2016).
44. Y. Xiao, N. Jing, and X. Li-Jost, Uncertainty under quantum measures and quantum memory, *Quantum Inf. Process.* **16**, 104 (2017).

4 Prologue to Quantum Computing and Blockchain Technology

Manasa C.M., Pavithra N.
Presidency University

CONTENTS

4.1 INTRODUCTION TO QUANTUM COMPUTING AND BLOCKCHAIN TECHNOLOGY

This section introduces the various aspects of Quantum Computing and BC technology. The details are presented as follows.

4.1.1 QUANTUM COMPUTING

Today's computers are based on classical physics, which limits their capabilities when it comes to operating efficiently. However, they can also behave very differently when observed in real-world conditions. A quantum system can have multiple states at the same time, which can affect its interference effects. Also, systems that are spatially separated may have non-local effects. Quantum computing is an

DOI: 10.1201/9781003250357-4

advanced method of processing that depends on the philosophy of quantum mechanics and inconceivable marvels. Quantum computing is a delightful blend of physical science, math, computer science, and data hypothesis. As a more comprehensive framework of material science than traditional mechanics, quantum mechanics leads to a more comprehensive model of preparing quantum processing, which may potentially deal with issues that conventional computers cannot address. They have their own quantum bits, also known as "Qubits", to store and control information, as opposed to traditional computers, which rely on old-fashioned handling that uses twofold pieces 0 and 1 freely. Quantum computers are computers that use this type of processing [1]. Circuits with semiconductors, logic entryways, and integrated circuits are impractical in such small computers. As a result, it employs subatomic particles such as molecules, electrons, photons, and particles as their constituents, as well as their data of twists and states. They can be combined and superimposed. As a result, they can process in equal utilizing memory proficiently, which is all the more remarkable. Quantum computing is the dominant approach capable of defying the Turing [2] hypothesis. As a result, quantum computers outperform traditional computers. Here, we present a prologue to the major ideas and a few thoughts on quantum computing.

Quantum computers can tackle computation issues that old-style PC can handle. As indicated by the Church-Turing [2] theory, the opposite was additionally evident that traditional computers also tackle any issues of quantum computers as well. It implies that they give no additional advantage over old-style computers as far as calculability is concerned, yet there are some intricate and incomprehensible issues that cannot be tackled by the present ordinary computers in the real world. It requires additional processing power. Quantum computers can take care of any issues in lesser time, known as "Quantum Supremacy".

Shor [2] has expressed that these computers also assist with addressing those issues impressively proficiently like in seconds without being heated more. He created calculations for considering enormous numbers rapidly. As estimations depend on the likelihood of a particle's current state before it is really known. These possess the capacity to deal with information in a dramatically immense amount. It likewise clarifies, cryptographic keys can be broken quickly by a quantum computer and this allows an eavesdropper to listen into private communications and pretend to be someone whom they are not. Quantum computers accomplish this by quickly reverse calculating or guessing secret cryptographic keys, a task that is considered very hard and improbable for a conventional computer. It can uncover private and ensure restricted data. In any case, the upsides of quantum computers are likewise remembered that is altogether more than their imperfections. Consequently, further exploration is going towards a more promising time to come.

While developing the regular PC, it was remembered that semiconductors' execution, particularly while getting more modest, will be influenced by commotion if any kind of quantum phenomenon happens. They attempted to stay away from quantum wonders totally for their circuits. However, the quantum PC adjusts an alternate method rather than utilizing traditional pieces and even chips away at the quantum theory itself. These computers utilize qubits which are undifferentiated from old-style bits and consist of two states (either 0 or 1) with the exception that it ensures few quanta properties where it can consist of the two qualities at the same time prompting an idea of superposed bits.

The qubit is the fundamental unit of quantum data, addressing elementary particles such as atoms, electrons, and so on like a computer's memory, while their control mechanism functions similarly to that of a PC's processor. It is either zero, one, or both at the same time. It is multiple times more impressive than the present most grounded supercomputers. Creation and the execution of qubits are gigantic difficulties in the area of designing. They obtain both advanced and simple characteristics that also give the quantum PC its supercomputing power. Their straightforward phenomenon demonstrates that quantum entryways have had no commotion limit, and its advanced phenomenon provides a way to compensate for this genuine shortcoming. In this way, the methodology of rationale doors and reflections made for traditional computing devices is of no utilization in quantum computing.

4.1.2 BC Technology

- Blockchain (BC), the establishment of Bitcoin, has got a wide range of considerations recently. Recently, digital money has become a popular expression in both industry and academia. Bitcoin exchanges take place without the involvement of a third party due to an exceptionally well-planned information stockpiling structure. Each square in the chain contains a different exchange, and each time another exchange takes place within the BC, a record of that exchange is appended to the record of every member. The term Distributed-Ledger Technology (DLT) points to a decentralized data set that is managed by a group of people. BC was presented through Bitcoin. In a distributed environment, transactions between two or more parties cannot be carried out without centralized authority; BC conquers the single point of failure problem which is introduced by the use of central authority. The innovation in building Bitcoin is a cryptocurrency that was first proposed in 2008 and implemented in 2009 [3]. Decentralization, persistence, namelessness, and auditability are the key characteristics of BC innovation. With these characteristics, BC can significantly reduce costs while increasing effectiveness. Businesses that require consistent quality and trustworthiness can use BC to attract customers. Furthermore, BC is dispersed and can avoid a single weak spot situation. The classifications of BC technology are as follows: Private technology and public technology. The main distinction is the job of the client in the organization and how the characteristic is observed. In a private BC, the maker of the organization knows from the start who the members are. In a public organization, you cannot fabricate a consent-based arrangement and the clients have all certifications of anonymity. It has grown quickly and can be utilized in numerous uses of different spaces like decentralized methodologies. BC innovation has become one of the most well-known procedures that will change the world, for the most part, due to its few elements like decentralization, permanence, and peer-to-peer exchanges. It gives a successful and lucid answer to some issues. Principally, BC is a decentralized appropriated data set innovation secure by means of cryptography calculations. This record data set is an affix data set; also, it cannot be adjusted or, on the other hand, it cannot be changed. BC functions on a peer-to-peer organization, each hub on this network contains an indistinguishable duplicate of the BC. Complete exchanges and timestamps are registered quickly, at that point, shared without the need of any outsider confided in power. A BC comprises of different sorts of parts, each with a particular job to carry out in making a BC network [4].

1. **Ledger:** A BC record is an appropriated, unchanging record of the historical backdrop of the BC, used as decentralized information capacity.
2. **Peer-to-Peer Network:** The record is put away, refreshed, and kept up with, utilizing a peer to peer organization, each hub consists of a duplicate of the record. Hubs add to every record update by following an agreement convention.
3. **Membership Services:** Few sorts of BC expect approval to join. For sure, enrolment administrations verified, approved, and dealt with the personality of clients on the BC.
4. **Smart Contract:** It refers to programs that sudden spike in demand for the BC, considering a bunch of conditions defined earlier. Clients can interface with smart contract likewise that they communicate with programs on a standard PC. Plus, it additionally gives security and is put away the computerized record.
5. **Wallet:** It stores the client's qualifications and keeps up key sets of the client to make and sign exchanges utilizing a digital signature.
6. **Event:** It refers to notices of updates and activities on the BC, instances of occasions: expansion of another block to the BC, creation, and scattering of another exchange across the peer-to-peer organization, and warnings from SCs on BCs that help such agreements.
7. **System Management:** BC is an advancing framework that should change to address the issues of its clients. Without a doubt, frameworks offers the executives, that the chance of making, changing, and observing BC parts to address the issues of its clients.
8. **Frameworks Integration:** As BC has created and expanded in usefulness, it has become more normal to incorporate BCs with other outside frameworks, generally with the utilization of SCs. While this is anything but a particular component of the BC, frameworks mix is incorporated to perceive this ability.

4.2   HOW DOES QUANTUM COMPUTER WORK?

4.2.1   The Architecture of Quantum Computers

Design can be thought of as an outline. The design of the quantum PC is a hybrid of traditional and quantum components, and it is divided into five layers, shown in Figure 4.1, with each layer serving as the PC's utilitarian component.

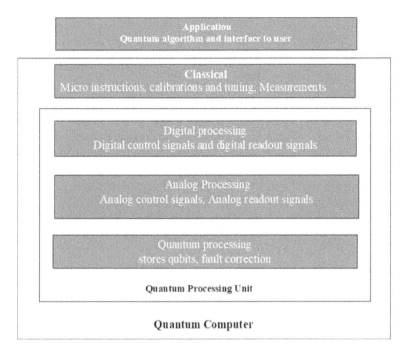

FIGURE 4.1 The architecture of quantum computers

- **Application Layer:** This layer does not form part of a quantum computer. It has been utilized to address user interface, which is the working framework for a quantum PC, programming environment, and other components needed to calculate reasonable quantum calculations. It does not require any equipment.
- **Classical Layer:** It improves and collects quantum calculations into micro-instructions. This also measures quantum state estimation restored from equipment inside the lower layers and feeds it into a conventional calculation to produce the result.
- **Digital Layer:** It converts microinstructions into the messages (beats) required by qubits, which behave like quantum logic gates. This is the digitized representation of the analogue pulses required in the sublayers. Furthermore, as a response to the preceding classical layer, it provides quantum estimation in order to consolidate quantum results to the end product.
- **Analog Layer:** It generates voltage signals with stage and amplitude modulations, similar to waveforms, for transmission to the lower layer, where quit functions are also executed.
- **Quantum Layer:** On a single chip, it is combined with the analogue and digital layers. It is kept at room temperature and is used to store qubits. This section is about fault correction. This component has an impact on how the computer performs better.

The Quantum Processing Unit (QPU) is made up of three layers: digital processing, analogue processing, and quantum processing. The QPU and the classical layer comprise the quantum computer. Both the digital and analogue layers work at room temperature.

4.2.2 WORKING OF BLOCKCHAIN

A BC is the next level of data management that is organized into a connected rundown of squares. Every square has two sections: a header and a body. The square header consists of metadata like Merkle tree root, nonce, timestamp, and some more. Though the square frame comprises a bunch of related exchanges, as displayed in Figure 4.2 below.

A square header is a hash of numerous objects dictated by BC, yet regularly comprises of the previous block header, the Merkle foundation of the current square, and the timestamp.

By counting the previous block header hash, blocks are connected "affixed" together (see Figure 4.3); this guarantees the trustworthiness since transforming one past square will require evolving each past block. BC engineering comprises of six primary parts incorporated into various layers establishing the BC, as displayed in Figure

- **Data Layer:** It contains the tied information blocks and the connected procedures including deviated encryption, timestamp innovations, hash calculations, and Merkle trees. Every hub uses the hash work, public-key plan, Merkle tree, etc., in request to embody exchanges and the code got into a new square with a timestamp demonstrating the hour of the creation of that square. Then, at that point, this new square will interface with the principal block to be added to the chain as a new block.
- **Network Layer:** It contains a dispersed organization instrument, specifically, information transmission and confirmation components. The layer gives every hub the investment in the exchange enlistment and information check measure.

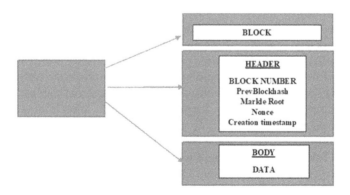

FIGURE 4.2   BC square

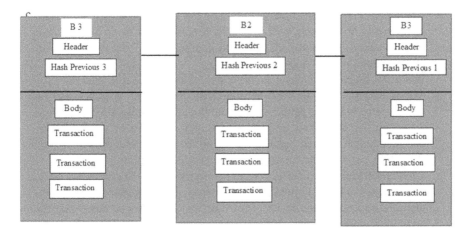

FIGURE 4.3 Square header of blockchain

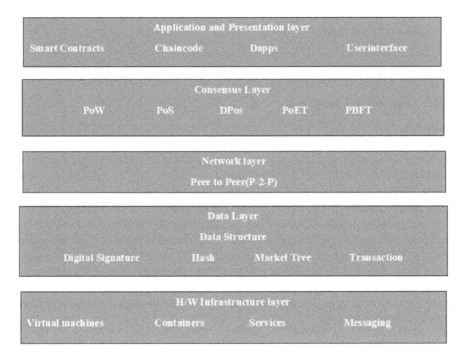

FIGURE 4.4 Layered architecture of blockchain [4]

- **Consensus Layer:** This layer is a gathering of different sorts of agreement components, to keep up with information consistency in an appropriated network and to guarantee adaptation to internal failure of a common vault between appropriated hubs.
- **Incentive Layer:** This layer incorporates monetary awards into BCs, the money system, and the cash appropriation instrument to guarantee the greatest advantage for people who create the following square. For sure, to propel the organization to proceed with its endeavours in information check and accomplish the well-being of the whole chain-of-blocks framework, the excavators get monetary motivations in light of their commitments.
- **Contract Layer:** The contract layer is concerned with the contract in question. Because a badly written or implemented contract layer can have financial consequences, significant effort must be taken to guarantee that the contract is issued appropriately and without potential flaws. The programme must also be verifiable, secure, and dependable. In addition, the contract must be correctly implemented and free of potential flaws. Verifiable, secure, and dependable software is required.
- **Application Layer:** This layer incorporates probability applications, situations, and use cases.

4.3   APPLICATION AREAS OF QUANTUM COMPUTING AND BC

4.3.1   QUANTUM COMPUTING

Numerous quantum algorithms are developed for quantum computers which accelerate speed that is a consequence of a few essential numerical strategies like Fourier change, Hamiltonian re-enactment, and so on. Most calculations require an enormous qubit of the best quality and few mistake rectification to give valuable functionalities. Instead of general joined applications, these algorithms are shaped in blocks since joined applications are not feasible. Subsequently, it is an extraordinary test to make quantum applications that are actually basically helpful alongside giving speed with no mistake. The beneficial use of a quantum computer's potential utilities is an area of active research. It is anticipated that these applications will require lesser qubits and will be accomplished with fewer codes. Because of the properties of qubits, it is possible that calculations on quantum computers will run faster. From cyber security to drug exploration here are a few different ways quantum will work with significant headways.

1. **Cyber Security and Cryptography:** The web-based security space is currently very dangerous due to the increasing number of security attacks occurring across the globe on a consistent basis. Despite the fact that organizations are putting in place critical security systems in their organizations, the operation becomes intimidating and illogical for old-style digital computers. As a consequence, network security has remained a major concern all over the world. With our increasing reliance on technology, humans are becoming significantly more vulnerable to threats. Quantum computing, in

tandem with Artificial Intelligence, can aid in the development of many strategies to combat these network security threats. Furthermore, quantum computing can aid in the improvement of encryption techniques known as quantum cryptography.

2. **Drug Development:** A very difficult task in quantum computing is the development of medication. Drugs are created through the experimentation strategy, which is not only costly but also a hazardous and time-consuming task to complete. Scientists believe quantum computing is a used tool for understanding medications and their interactions with humans, saving drug companies significant money and time. These advancements in processing could significantly improve effectiveness by allowing organizations complete additional medication disclosure in order to discover innovative clinical drugs for a more prosperous pharmaceutical industry.

3. **Financial Modelling:** It is critical for a finance-based industry to find the right mix of investments based on expected returns, risk, and other factors in order to survive. To accomplish the method of 'Monte Carlo', reproductions consistently run on customary computers, which burn through a massive measure of computer time. Finance professionals are in charge of billions of dollars, and even a tiny increase in the regular return can be extremely beneficial to them. Organizations can work on the character of the arrangements as well as reduce the opportunity to promote them by using quantum technology for complex computations.

4. **Weather Forecasting and Climate Change:** At the moment, the process of analysing weather conditions by traditional computers can take more than the actual climate to change. Yet, quantum computers have the power to crunch immense measures of information, in a short period, surely giving rise to weather system displaying and allowing researchers to predict the change in climatic designs in a matter of moments and with amazing precision — something that could be critical for the passage of time when the globe changes owing to environmental changes. Weather forecasting is difficult to anticipate because it takes into consideration a variety of elements such as temperature, air pressure, and density. Quantum AI can help with example recognition, making it easier for academics to anticipate extreme weather events and perhaps saving thousands of lives every year. Weather forecasters can also use quantum computers to create and analyse more detailed environmental models, which will help them better understand environmental change and how to mitigate it.

5. **Artificial Intelligence:** ML and AI are some of the dominant areas at this moment, as the rising advancements have entered each and every aspect of people's lives. As a part, limitless applications are seen every day in the recognition of voice, sound, and handwriting. In most situations, as the amount of usage expanded, it is converted into a complicated assignment for customary computers, to coordinate the exactness and speed. Also, that is the environment where quantum computing can help in handling difficult issues in an extremely short time, which would have required customary computers for a greater number of years.

4.3.2 BC

BC is a type of database that is non-centralized, dependable, and hard to use for fake purposes. Bitcoin is only one of the monetary applications that utilize BC innovation; there are likewise others like keen agreement and hyper ledger. BC innovation can, in this way, be utilized to make numerous applications.

4.3.2.1 Commercial Applications

- **Bitcoin:** Bitcoin (or digitized money) was first presented by Satoshi Nakamoto in 2008. Across the shared organization, Bitcoin utilizes a BC public record to make exchanges. Instances of dynamic Bitcoins are BTC Jam, Bit bond, BitnPlay, Codios, and DeBuNe.
- **Ripple:** The real time gross settlement framework employs ripple for trade and settlement across a distributed organisation, a decentralized trade that focuses on the financial market. Billion, Coinbase, Bites, Kraken and Crypto Sigma, and Stellar are some other well-known cash trade and settlement frameworks.

4.3.2.2 Non-commercial Applications

- **Ethereum:** Vitalik Buterin, a digital money scientist, created a next-generation smart contract and decentralized application platform. It makes use of a BC-based suitable distributed computing system with a Turing complete scripting language that allows clever agreements to be handled on it.
- **Hyperledger:** The Hyperledger project, founded by Linux, creates BC innovations for work, assisting enlisted individuals. Hyperledger is an open-source collaborative effort created to propel cross-industry BC inventions. The Linux Foundation facilitated a worldwide coordinated effort, which recalls pioneers for finance, banking, supply chain, the Internet of Things, fabrication, and innovations. BC with its attributes recorded above is thought of adaptable and appropriate to replace numerous applications. Many use cases are currently investigated in various fields other than monetary and payments, including portfolio management reporting, product distribution, security and anti-fraud measures.

The three BC executions utilized are discussed as follows:

i. The Public BC is the most well-known methodology, for example, Bitcoin where anyone can send exchanges and hope to see them in the following square of the chain. Members are engaged with the confirmation cycle "agreement" and help with figuring out which exchanges get added into the record and which are not.

ii. Consortium BC is viewed half arrangement of public BC and private BC; it allows certain hubs to take part in the Consensus cycle. For instance, a central bank allows only trusted banks to provide the fundamental controls, confirming trades before they are added to the square. Furthermore, the record's read consent might be restricted. Similarly, this type is frequently referred to as a partially decentralised BC.

iii. The Private BC type comprises of one association that has the authorization to make (compose) new exchanges anyway the peruse authorizations are confined to just chosen hubs. It is presumably utilized by the board/reviewers' organizations that need to control a few touchy exercises which are estimated inside, where public read admittance doesn't make a difference inside their applications.

4.3.3 Challenges of BC

The interchanges are expanded pace by pace, the BC becomes unmanageable. Every hub needs to store all the interchanges happened, to accept them on the BC in fact that they need to check if the source of the current interchange is unspent or not. Moreover, because of the initial limitation of square size and the time taken to produce another square, the Bitcoin BC can just handle almost seven interchanges every second, which cannot satisfy the necessity of handling a huge interchange continuously. In the meantime, the limit of blocks is tiny, many little interchanges may be postponed since diggers incline toward those interchanges with high expenses.

There are many endeavours proposed to address the various issues of BC, which could be sorted as follows:

- **Storage Improvement of BC:** Formerly, it was very difficult for cluster to work complete duplication of record. Bruce proposed a novel e-cash scheme in which the organisation disposes of old exchange records. All non-empty locations are saved by using "account tree" by which steadiness is maintained. Otherwise, a customer could assist with fixing this issue [6]. Ver Sum permits lightweight customers to re-evaluate costly calculations across massive information sources. It ensures that the computation result is correct by comparing results from various servers.
- **Redesigning BC:** The term "Bitcoin-NG" (Next Generation) was coined. Bit-coin-preliminary NG's concept is to divide a normal square into two sections: key square for pioneer political race and micro block for storing interchanges. The time is divided into epochs according to the convention. Diggers must hash to produce a key square in each iteration. When the key block is constructed, a hub is converted into an explorer, which is responsible for the creation of micro blocks. Bitcoin-NG too expanded the massive (longest) chain procedure in which micro blocks convey no weight. Thus, BC is maintained and the trade-off between block size and network security has been tended to.
- **Safety Leakage:** BC can save a specific measure of protection by utilizing the public key and private key, and the public key assists clients in executing, with almost no authentic character openness. Notwithstanding, BC cannot ensure conditional safety considering the upsides of entire interchanges and balances for each open key are easily noticeable. Also, a new review has

shown that a client's Bitcoin interchanges can be brought together to bare client's data. When clients are behind Network Address Translation (NAT) or firewalls, each customer can be recognized with a group of hubs associated with, this set can be learned and utilized to discover the beginning of an exchange happened. Various techniques have been proposed to further develop the security of BC.

- **Mixing:** Client addresses are fictitious in the blockchain. However, it is possible to link addresses to clients in a genuine way because many clients make interchanges with very similar addresses as frequently as possible. By moving assets from different data locations to different addresses, blending administration creates anonymity. Client Ally with address A must send assets to Bobby with address B for model. The relationship between Ally and Bobby could be disclosed if Ally just exchanges input and yield addresses. As a result, Ally could send assets to Carol, a presumed middleman. Carol then transfers assets to Bob using various sources of information c1, c2, c3, and so on, as well as various yields d1, d2, B, d3, and so on; Sway's address B is likewise contained in the yield addresses. As a result, it becomes more eager to discover a connection between Ally and Sway. Nonetheless, the middleman could be deceptive and intentionally reveal Ally and Bob's personal information. Rather than Bobby's location, it's also possible that Carol moves Ally's assets to her own location. Mixcoin provides a straightforward strategy for avoiding unethical practices. With its private key, the mediator scrambles clients' requirements, including reserve sum and move date. Then, if the intermediary did not move the currency, anyone could verify that the middleman cheated. Nevertheless, theft is acknowledged at the same time. Conjoin relies on a central blending server to rearrange yield locations in order to prevent robbery. Furthermore, influenced by Coin Shuffle, conjoin uses decoding mix nets for address reposition.

- **Anonymous:** In Zero coin, the Zero-information confirmation method is used. Diggers are not required to approve an exchange with a computerized signature, but they must approve coins that have a place on a list of significant coins. To avoid exchange diagram examinations, the starting point of each instalment is unlinked from exchanges. In any case, it reveals the objective and sums of the instalments. To resolve this issue, no money was proposed. Non-intuitive variation of knowledge is used in Zero cash, Zero-information briefs. The exchange amounts and the upsides of coins held by clients are concealed.

4.3.4   IMPORTANCE OF QUANTUM COMPUTING BLOCKCHAIN FOR QUANTUM SATELLITES AND DRONES

In the future, the incorporation of quantum computing technology, AI, drones, and security considerations provides a plethora of applications [7,8]. Quantum sensors coupled with drones, unmanned underwater vehicles (UUVs), unnamed aerial vehicle system (UAVs), or unnamed surface vehicles (USVs) can aid with a variety of applications, including collision detection in delivery-based systems, human-made structures, spectrum measurements, and more [9]. As discussed by Choi [10] in the development of the first "quantum drone" for inaccessible air-to-ground data communication, a kilometre-long drone-based quantum network prototype that effectively transmitted a quantum signal [11], drones can also be used to distribute quantum keys. This experiment demonstrates quantum communication's ability to ensure safe message interchange. Two users can share a pair of entangled photons with a distinct mechanical relationship in this space. Quantum key distribution via quantum satellites occurs at a faster rate. Similarly, quantum satellites have been used to develop multimedia communications (audio, video, and text).

In many different areas, blockchain is implemented such as drone air traffic management, drone insurance, fleet management, identity management, and data

---

TABLE 4.1
Advantages and Disadvantages of Quantum Computing

Advantages	Disadvantages
1. Quantum computers are capable of solving complex numerical issues which customary computers find difficult to solve in a real-time period.	1. New algorithm must be written for each type of calculation. Quantum computers cannot operate in the same way that traditional computers do; they require specific algorithms to complete tasks in their environment.
2. It gives computing capability which can adequately measure a huge volume of data (2.5 Exabyte day by day equivalent to 5 million computers) generated from all over the world to extract insight from it.	2. Because of their high cost, they are not widely available. Furthermore, due to the fact that these computers are still in development, they have a high rate of error. Quantum computers work well with 10 qubits, but their accuracy decreases as the number of quantum bits increases, such as to seven quantum bits.
3. Quantum computers work parallelly and utilize less measured power, hence, diminishing the force utilization up to 100 to multiple times due to the quantum tunnelling [2].	3. It is extremely fragile and prone to errors. Any type of vibration has an effect on subatomic particles such as electrons and atoms. As a result, failures, flaws, and destruction are all possible. It causes "decoherence", which demonstrates a lack of logic in quantum mechanics.
4. A quantum computer in its entirety is "a large number of times" quicker than any conventional PC. Google has created a quantum computer in its lab that is a hundred times quicker than any regular PC.	4. As the processing in these computers is done very deeply so it needs a temperature of negative 460 degrees F. This is the lowest temperature of the universe and it is very difficult to maintain that temperature.
5. These computers can make high encryption and is good at cryptography. It is impossible to break the security of quantum computers. Recently China has launched a satellite that uses quantum computing and china claimed that this satellite cannot be hacked.	5. The current Internet of Things (IoT) would be rendered insecure due to advances in quantum computing. Cryptographic procedures, government and private massive organizations' databases, banks, and security frameworks can all be hacked.

security. Drones have no way of identifying themselves while flying; as a result, they are able to fly close to businesses and exploit wireless network vulnerabilities without being detected. Drone identification would help detect and monitor rogue drones. Data pertaining to drone identity will be recorded in the form of blocks in each node of a blockchain network [12]. The identity of a drone joining such a network will be confirmed by a consensus of multiple drones. Only authorized drones will be allowed to join a network. Drones that have been compromised will be unable to communicate with other drones and gadgets on the network. The Blockchain-AI supervision scheme [7], in which a batch of drones included by ID is set with AI and made busy in controlling crisis framework autonomously. Drones are used in a variety of fields, including medicine and commerce, which necessitates security. Shallow BCs will act in distant areas, and a two-phase shallow protection mechanism is used to handle multiple operations. Experiments were conducted in order to validate the system. Despite the fact that investigation requires all real-world applications, administration tasks were completed satisfactorily. Drone-based data collection and AI model training service providers are becoming increasingly popular. Data sharing between individually operated UAVs, on the other hand, is difficult due to strict data security regulations [13].

4.4   CONCLUSION

Quantum computing is a cutting-edge processing method based on quantum physics principles and incomprehensible miracles. With its main qualities of decentralization, persistency, anonymity, and auditability, the blockchain has proved its potential to revolutionize traditional industries. This work has addressed the applications, security challenges, and importance of adopting quantum computing and blockchain technology in drones and satellites. Quantum computing and blockchain are expected to have enormous capability in near future. This work will be advantageous in diverse directions of applications using quantum computing and blockchain [14,15].

REFERENCES

1. R. Versluis, "Here's a blueprint for a practical quantum computer," IEEE Spectrum, 2020, https://spectrum.ieee.org/computing/hardware/heres-a-blueprint-for-apractical-quantum-computer.
2. D. Nield, "Google's quantum computer is 100 million times faster than your laptop," Science Alert, 2015, https://www.sciencealert.com/google-s-quantum-computer-is100-million-times-faster-than-yourlaptop.
3. Z. Zheng, S. Xie, H. Dai, X. Chen, and H. Wang, "An overview of blockchain technology: Architecture, consensus, and future trends," In *2017 IEEE International Congress on Big Data (BigData Congress)*, pp. 557–564. IEEE, 2017, doi: 10.1109/BigDataCong ress.2017.85.
4. M. Fartitchou, K. El Makkaoui, N. Kannouf, and Z. El Allali, "Security on blockchain technology," In *2020 3rd International Conference on Advanced Communication Technologies and Networking (CommNet)*, pp. 1–7. IEEE, 2020.

5. R. Barends, J. Kelly, A. Megrant, A. Veitia, D. Sank, E. Jeffrey, T.C. White, J. Mutus, A.G. Fowler, B. Campbell, and Y. Chen, "Logic gates at the surface code threshold: Supercomputing qubits poised for fault tolerant quantum computing," *Nature*, vol. 508, pp. 500–503, 2014.

6. P. Tasatanattakool and C. Techapanupreeda, "Blockchain: Challenges and applications," In *2018 International Conference on Information Networking (ICOIN)*, pp. 473–475. IEEE, 2018, doi: 10.1109/ICOIN.2018.8343163

7. A. Islam, T. Rahim, M. Masuduzzaman, and S. Y. Shin, "A blockchain-based articial intelligence-empowered contagious pandemic situation supervision scheme using Internet of Drone Things," *IEEE Wireless Commun.*, 2021, doi: 10.1109/MWC.001.2000429

8. H. Houssein, A. E. Samhat, M. Chamoun, H. El Ghor, and A. Serrhouchni, "On blockchain technology: Overview of bitcoin and future insights," In *2018 IEEE International Multidisciplinary Conference on Engineering Technology (IMCET)*, pp. 1–8. IEEE, 2018, doi: 10.1109/IMCET.2018.8603029

9. S. P. Jordan and Y.-K. Liu, "Quantum cryptanalysis: Shor, grover, and beyond," *IEEE Secur. Privacy*, vol. 16, no. 5, pp. 14–21, 2018, doi: 10.1109/MSP.2018.3761719

10. W. Y. B. Lim, J. Huang, Z. Xiong, J. Kang, D. Niyato, X.-S. Hua, C. Leung, and C. Miao, "Towards federated learning in UAV-enabled internet of vehicles: A multidimensional contract-matching approach," *IEEE Trans. Intell. Transp. Syst.*, vol. 22, no. 8, pp. 5140–5154, 2021, doi: 10.1109/TITS.2021.3056341

11. A. D. Hill, J. Chapman, K. Herndon, C. Chopp, D. J. Gauthier, and P. Kwiat, "Drone-based quantum key distribution," *Urbana*, vol. 51, pp. 61801–63003, 2017.

12. A. Ossamah, "Blockchain as a solution to drone cybersecurity," In *2020 IEEE 6th World Forum on Internet of Things (WF-IoT)*, pp. 1–9. IEEE, 2020, doi: 10.1109/WF-IoT48130.2020.9221466

13. A. Kumar, S. Bhatia, K. Kaushik, S. M. Gayathri, S. G. Devi, A. D. Diego, and A. Mashat, "Survey of promising technologies for quantum drones and networks," *IEEE Access*, vol. 9, pp. 125868–125911, 2021, doi: 10.1109/ACCESS.2021.3109816

14. M. Schirber, "Quantum drones take light," *Physics*, vol. 14, p. 7, 2021, doi: 10.1103/PHYSICS.14.7

15. C. Q. Choi, "World's first quantum drone for impenetrable air-to-ground data links takes off," 2019. Accessed: Jun. 25, 2021. [Online]. Available: https://spectrum.ieee.org/tech-talk/computing/networks/quantum-drone

5 Quantum Computing Application for Satellites and Satellite Image Processing

Ajay Kumar and B.S. Tewari
G.B. Pant Institute of Engineering and Technology

Kamal Pandey
Indian Institute of Remote Sensing

CONTENTS

5.1 INTRODUCTION

In space research, a simple approach to quantum approaches has been pioneered by a team headed by Paolo Villoresi at the University of Padua of Italy. The technique consists of the attachment of reflectors and other basic equipment to regular satellites, which is quite inexpensive. Villoresi et al. (2008) discovered that photons bouncing back to earth from an existing satellite preserved their quantum states and were received with error rates low enough to be used in quantum cryptography. The current use of satellites to obtain information about earth's features by sensing the radiation reflected/emitted by various objects on the earth's surface has the potential to solve a wide range of real-world problems, including natural resource management, disaster management, town planning, precision agriculture, and so on. Satellites were originally designed for defence purposes only, but scientists and researchers soon discovered a wide range of applications for satellites, and they have now become an integral part of human life, playing an important role in a variety of sectors such as communication, navigation, and weather monitoring. These satellites orbit the planet indefinitely, collecting photographs of the surface and generating massive amounts of data. According to many researchers, quantum computers will be much faster than current supercomputers. The analysis of this vast amount of data using classical computers is a time-consuming operation that delays real-time monitoring of occurrences on the earth's surface.

Quantum computers will be extremely useful in running various models established for natural disasters (floods, fires, earthquakes, and so on) and providing conclusions for further decision-making (Aoki & Shimokawabe, 2013). The quantum computer will also aid in the improvement of weather forecasting models, as current computing takes a long time to produce findings (Mastriani et al., 2021). Weather forecasting is a complicated process that necessitates a large number of inputs from satellite meteorological stations (Zhang et al., 2021). To run a weather forecasting model, it must cover a large portion of the earth's atmosphere, which is then divided into smaller grids (Matuschek & Matzarakis, 2010). As a result, with current computational capabilities, generating data from weather forecasting models takes a long time. Quantum computing is also anticipated to help Artificial Neural Network (ANN) and Machine Learning (ML) techniques extract relevant information from enormous datasets. In this chapter, we look at how quantum computation can be used to improve current satellite data processing techniques, as well as their applications in weather forecasting and real-time monitoring of natural disasters.

5.2 SATELLITES

A satellite is a man-made vehicle or platform that orbits the earth in a fixed orbit in space and carries application-specific sensors and instruments. Satellites are classified based on the payload they carry and the orbit they are placed in. For example, communication satellites must continuously view the earth's surface; thus, they are placed in the geostationary orbit at a height of 36,000 km above the surface (Lillesand, 2016). Similarly, remote-sensing satellites are equipped with imaging sensors that capture photographs of the earth's surface in several electromagnetic bands. The recorded data is sent to a ground control station where it is processed and disseminated for specialised applications such as land use and landcover dynamics, crop yield forecasting, hydrological modelling, and urban dynamics, among others.

The modern world makes use of satellite technologies for a large number of societal applications, i.e., internet, telecommunication, weather prediction, etc. Satellites have significantly contributed to the growth of other technologies as well in the last few decades. It provides a quick mechanism for dissemination of information across the globe.

Newton's Cannonball thought experiment, which was described in *Philosophi Naturalis Principia Mathematica*, was the first to introduce the notion of an artificial satellite (1687) (Smith, 2008). Later, Exploring Space Using Jet Propulsion Devices was written in 1903 by Konstantin Tsiolkovsky (1857–1935) and was the first academic book on the use of rocketry to launch spacecraft. Tsiolkovsky was born in Russia and died in 1935. The orbital speed necessary for a minimum orbit was calculated, and he determined that it could be attained by a multi-stage liquid propellant-fuelled rocket with many stages. In 1957, the Soviet Union launched Sputnik 1, the world's first artificial satellite, as part of the Sputnik programme, with Sergei Korolev acting as the spacecraft's principal designer. With the help of Sputnik 1, scientists were able to gather information on radio-signal dispersion in the ionosphere; and by analysing its orbital fluctuation, they were able to determine the density of upper atmospheric layers. The unexpected success of Sputnik 1 sparked the Sputnik crisis in the United States, reigniting the Cold War and launching a space race that would last for decades (Burrows, 1999). Satellites could be classified based upon their orbit, altitude, and application, as shown in Figure 5.1. Satellites are classified into three categories based on orbit. Satellite revolving in the sun-synchronous is designed in such a way that it always captures any particular place of earth on fixed local sun time. Most of the remote-sensing/earth-observing satellites are sun-synchronous. Images captured by sun-synchronous orbit satellites are excellent for studying temporal changes. Sun-synchronous orbits are also polar orbits, but all the polar

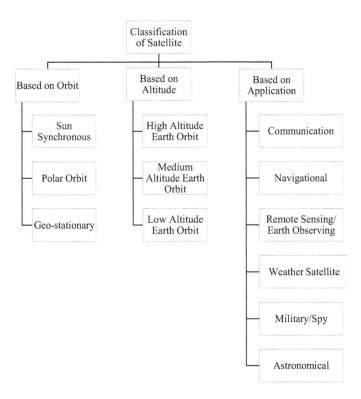

FIGURE 5.1   Classification of satellites

orbits may not be sun-synchronous. Geostationary satellites are those satellites which revolve in the equatorial orbit of the earth and are synchronised in such a way that they remain above a certain fixed area of the earth. All communication and weather satellites are actually geostationary satellites.

Based on the altitude, there are again three categories of satellite—high-, medium-, and low-altitude earth orbit satellites. All the communication and weather satellites revolving in the geostationary orbit of the earth are high-altitude earth orbit satellites. Their altitude is approximately 36,000 km above the earth's surface. All the satellites that constitute the space constellation of Global Navigation Satellite Systems (GNSS) are actually medium-altitude earth orbit satellites. Their altitude is approximately 20,200 km above the earth's surface (Lillesand, 2016). Low-altitude earth orbit satellites are those that orbit at a height of less than 2,000 km. Satellites can also be classified according to their uses, such as communication, navigation, remote sensing/earth observation, weather satellites, military/spy satellites, and astronomical satellites. The life span of a typical satellite is relatively around 5–8 years only, which means a new satellite should be launched every 5–8 years in order to continue the services of the satellite. The major challenges faced in present satellite systems are listed below:

1. Integration and testing of satellite systems takes a long time.
2. The low life span of present satellites restricts the data collection for a longer duration.
3. There is a high volume of redundancy in data collected remote-sensing satellites.
4. Data security is also a major concern nowadays. The existing methods of security for data distribution are not foolproof.
5. The transmission of data using existing methods requires a large amount of power.
6. The amount of data generated by a satellite is huge, and it takes a lot of time to process the data using existing computing systems.

In the next section, we will discuss how quantum technologies will help in overcoming the above-mentioned issues.

5.3   QUANTUM TECHNOLOGIES

Quantum technologies are highly promising due to some unique features like quantum superposition and quantum entanglement. The quantum superposition allows multiple inputs in the form of qubits (|0> and |1>) which in turn reduces the inherent operations (computation process) to achieve output value. Due to the minimum number of operations in quantum computation, the speed of such quantum-based systems increases in an apparent manner.

5.3.1   Slow Process of Integration and Testing Satellite Systems

The existing methods of integration and testing of components of a satellite are very slow. The conventional process of manufacturing highly specialised and customised satellites is slow, which is one of the major drawbacks in the present scenario. The major reason behind this is that the existing systems lack the capability of solving the big numerical simulation problems at high speed. Alternatively, a fast quantum computer makes it possible to solve very complex simulation problems at a very high speed. The main inspiration for developing working quantum computers comes from quantum computers' ability to crack public-key encryption systems like the commonly used Ron Rivest, Adi Shamir, and Leonard Adleman (RSA) scheme (Rivest et al., 1978). Shor's quantum algorithm (1994) highlights the potential advantage of quantum computers in this sector by efficiently factoring huge integer numbers into prime factors in polynomial time, whereas the most efficient conventional techniques take sub-exponential time. Quantum computers will almost certainly be employed to solve complex scientific challenges once they are fully functional. Computer simulation technology is used to analyse difficult challenges that develop in several fields when computing scientific topics. In all simulation methodologies, the results' reliability is always an issue. The approximation results could be regarded as "the real answer," posing a societal risk. Therefore, we may trust the findings in cases where the approximations can be compared with measurements for a number of scenarios, but the results for difficult and new applications should be regarded with caution. Assume having a fully functional quantum computer, which is capable of solving the big complex numerical simulation problems in a few seconds. Obviously, it will result in speedy testing of satellite system components; thus, the process of developing a satellite will be faster.

5.3.2 Low Life Span of Satellite Systems

Because satellite systems have a limited life span, they are one of the major producers of space debris in earth's orbit. We humans have already done significant damage to the environment of the world in order to fulfil our own objectives. Sensors on the Department of Defense's global Space Surveillance Network (SSN) record approximately 27,000 bits of orbital debris, sometimes known as "space trash," according to the Department of Defense. Researchers have discovered that the near-earth space environment includes far more trash than previously thought. This material is too tiny to be spotted but big enough to pose a threat to human spaceflight and robotic missions. Because both the junk and the spacecraft are travelling at such high speeds (about 15,700 mph in low-earth orbit), even a little piece of orbital debris colliding with a spacecraft might result in catastrophic consequences for both. As previously stated in Section 5.1, we must deploy a separate satellite for each kind of application that we want to employ, as detailed in Section 5.1. A new satellite is launched by a space organisation funded by the United Nations every 5–6 years in order to continue to use the service provided by a specific satellite. Satellites have a limited life span; so, in order to continue to use the service provided by a specific satellite, a new satellite is launched every 5–6 years by a space organisation funded by the United Nations. The old satellite may provide the services but a new satellite is launched to ensure the continuity of the service. The problem of space debris can reduce significantly if the life span of the satellites could be increased by a great amount with the use of recent scientific advancements alongside quantum technologies. A satellite may become non-operational because of many reasons. Many times, a simple single component failure can result in a non-operational satellite, but power outages are the leading cause of satellite death. The majority of operational satellite systems produce only 7–12 kW. Next-generation designers, on the other hand, have begun to suggest large flexible floppy solar arrays capable of producing 50–60 kW. The use of gallium arsenide/germanium multi-junction cells, which have the potential to achieve solar cell efficiencies of more than 30%, is also being pursued for improved solar cell performance. Parallel efforts are being made to advance battery technology, particularly lithium-ion and fuel cell technology, in order to construct higher and higher-powered satellites. The remote sensing satellites consume more power in operations like processing the collected image data and transmitting it back to earth. When compared to the power needs of refrigerators and computers, the quantum processor, according to Colin Williams, D-Wave Systems' director of business development and strategic partnerships, consumes remarkably little power—only a fraction of a microwatt (Clerk et al., 2020). As a result, utilising quantum computing-based processing systems on a satellite would save a significant amount of power. If data is not broadcast to earth but is exchanged with earth stations via quantum entanglement, the data is not transmitted to earth. Theoretically, transporting data via quantum entanglement should use far less energy than communicating data via electromagnetic wave with the use of a transmitter to an earth station. These electromagnetic waves have to travel through the earth's atmosphere, where signals get attenuated because of scattering refraction and absorption. Many signals to noise ratios are very low which makes data useless. All these problems could be avoided with the application of quantum entanglement.

5.3.3 Redundancy in Collected Data by Remote-Sensing Satellites

Remote-sensing satellites are those satellites which continuously take the image of the earth's surface for various applications like meteorology, monitoring land use and natural resource management, disaster management, etc. An imaging satellite generates a huge amount of data in a day; and during the life span of an imaging satellite, the amount of data generated becomes exponentially huge. Most of the time, there is a lot of duplication in the data acquired, which means that many images include the same information as images collected in the past in the same location. Another explanation for the satellite image's redundancy is that it contains a number of bands, which means that the reflected energy received by the sensor is recorded in multiple wavelength bands. This is done in order to attain the goal of better object rearrangement based on reflectance in different wavelength bands. Popular remote-sensing satellite like Landsat has 12 bands; MODIS has 36 bands; and in hyperspectral remote sensing, the number of bands could be a few hundred. Increasing the number of bands in a satellite image leads to more and redundant information in the satellite image. Principal Component Analysis (PCA) is one of the common approaches that lower the dimensionality and redundancy of satellite pictures. Performing these processes on a massive dataset produced by satellites will require substantial processing resources. Hence, quantum computing in conjunction with AI and machine learning may make these activities quite smooth. Quantum physics is widely recognised for creating odd data patterns. Deep neural networks and other typical machine learning algorithms, as well as data-mining algorithms, can detect statistical patterns in data and produce data that has the same statistical patterns; in other words, they can recognise the patterns they make. In light of this revelation, the following ray of hope has been conceived. The ability of small quantum information processors to find patterns that are as difficult to perceive conventionally is achievable if they can produce statistical patterns that are computationally impossible for a conventional computer to construct.

5.3.4 Data Security—Quantum Key Distribution

It is an efficient method of classical encryption to utilise public-key cryptography, which employs different keys for encrypting and decrypting data. Decoding plaintext encrypted by the public key is only possible with the secret key that corresponds to the public key. The Rivest, Shamir, and Adleman (RSA) technique encrypts data by using the product of two big prime numbers, and it cannot be decrypted unless the prime factors are known in advance. With today's processing power, factoring is a tough problem to solve, which ensures the security of the system. The quantum Shor algorithm, on the other hand, has the capability of breaking the RSA cryptographic technique in a matter of seconds. From a practical viewpoint, once quantum computers with a big enough number of quantum bits are built, the RSA algorithm will be rendered ineffective. The fact that quantum key distribution (QKD) makes use of the quantum physics rule means that it will be impossible for a classical computer to defeat it (Bacsardi, 2013).

The exchange of quantum secret keys among users of a communication network permits users to send and receive messages in a secure manner, free of eavesdropping and data leakage. We are almost invulnerable if we utilise these keys in cryptographic protocols. Because of their computational complexity, traditional cryptosystems rely on algorithms to assure data security. However, because of the so-called Shor's factoring technique, a quantum computer is capable of cracking even such encryption (Liao et al., 2017). QKD techniques are a solution which uses public optical channels to securely distribute keys by exchanging quantum bits, which are then carried by single photons of light. The no-cloning theorem assures complete security—an eavesdropper's measurement of a quantum bit risks changing its state. This indicates that the eavesdropper is present. Quantum keys are distributed between two connecting users using photons that travel through optical fibres or through atmospheric line-of-sight channels. We can't expand quantum communication across oceans due to photon loss in the optical cable. A quantum communication network built on circling satellites, on the other hand, could cover the entire planet's surface.

Quantum computing is based on quantum phenomena and may be used to tackle a broad variety of issues, including prime factorisation, database searching, key distribution, and information coding, among others. It is presently being investigated and developed in more depth. Another big communication challenge is determining how to distribute a secret key among the many communication partners that are engaged in the process. This is one of the most significant issues in all of the communication, if not the most crucial one. In this case, the use of QKD is a secure approach that can be used if necessary. The multiple polarisations of photons that are used in QKD applications correlate to the distinct quantum states of matter. According to quantum physics, quantum bits (qubits) have a continuous space, and unitary operations on qubits likewise have a continuous space under the theory of quantum mechanics as well. Since the unitary quantum processes' accuracy is limited, it is required for them to be isolated from their surroundings in order to construct an idealised channel between them. This separation is important because the unitary quantum processes' precision is restricted. It is not feasible to accomplish total separation in principle since it is essential to interact with the qubits in order to complete the separation in real implementations. When it comes to successful free-space quantum communications, one of the most important needs is the capacity to consistently transmit quantum states in noisy settings, which is also one of the most difficult to accomplish. As a consequence of a shortage of effective quantum repeaters, optical fibre-based quantum key distribution has, according to current study, been restricted to a range of around 100 km. In contrast, the free-space quantum cryptography approach allows photons to be sent across great distances without affecting security.

5.4   QUANTUM COMPUTING FOR SATELLITE IMAGE PROCESSING

When it comes to space sector operations, quantum computing is a game-changing technology that can enhance operations by speeding up optimisation and machine learning processes. With the use of machine learning techniques, it is now feasible to automate picture categorisation in geographical data. New quantum algorithms provide one-of-a-kind answers to these issues, as well as the possibility of a future competitive edge over conventional approaches. When used in conjunction with Universal Quantum Computers, they enable the execution of completely generic quantum algorithms, with theoretically demonstrable speedups over conventional algorithms in specific situations.

The majority of the work involved in satellite image processing is divided into two stages. The first is pre-processing (Figure 5.2), and the second is post-processing (Figure 5.3). Georeferencing and geometric correction are the core functions of Stage 1. This is used for data restoration and rectification in order to compensate for data distortions produced by radiometric and geometric imperfections that are particular to the sensor and platform. The need for radiometric adjustments may arise as a result of changes in the scene light and viewing geometry, as well as changes in atmospheric conditions and sensor noise and responsiveness. Each of them is distinct in that it is based on the sensor and platform that was used to collect the data, as well as the current environmental circumstances. It may be essential to convert and calibrate the data to known (absolute) radiation or reflectance values in order to make data comparisons more straightforward. To compensate for differences in light and viewing geometry between images, models of the distance between the area of the earth's surface observed and its geometric connection with the satellite, the position of the sun, the sensor, and other factors may be developed (for optical sensors). This is required in order to more readily compare photographs taken by multiple sensors at different times or dates, or to mosaic many images taken by a single sensor while maintaining consistent lighting conditions from scene to scene. After the pre-processing of the satellite picture has been completed, the image will be subjected to post-processing. Image enhancement, classification, segmentation, feature

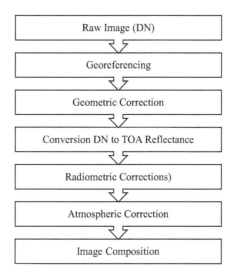

FIGURE 5.2 Steps in pre-processing of the satellite image

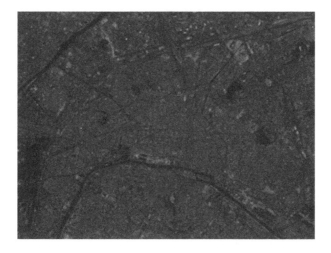

FIGURE 5.3 Unenhanced true colour composite

extraction, fusion, change detection, and compression are just a few of the processes that may be performed once data has been collected.

5.4.1 SATELLITE IMAGE PRE-PROCESSING

The images captured by sensors onboard satellite contain many errors, which must be removed before using them for any application. The primary and necessary corrective measures carried out in a satellite image are georeferencing and geometric correction.

Georeferencing means assigning an internal coordinate system to an image and relating it to the ground coordinate system. It also means that image is associated with locations in physical space. Geometric distortion is very common in satellite images due to the fact that satellites are revolving in the earth's orbit and taking pictures of the earth revolving about its own axis. Skewness caused by earth rotation is actually systematic in nature; so, it could be mathematically modelled and removed easily. Whereas skewness caused by a change in satellite position and velocity in space is random in nature, so difficult to remove/reduce it. In order to classify the satellite image, it must be converted back Top of Atmosphere (TOA) from Digital Number (DN). To view the image on a digital screen, the reflectance captured by the satellite sensor, i.e., TOA is converted into DN through a linear mathematical operation. The advantage of converting TOA to DN is better contrast and visibility in the satellite image. Now classification should be done on the TOA instead of DN for more accurate results (Lillesand, 2016).

The radiometric correction is used to adjust for lens corner, sun angle, and topographic relief-induced aberrations. Atmospheric adjustments are used to lessen the impact of local atmospheric variations on the amount of refraction, absorption, and scattering of radiation emitted or reflected by the earth's objects. For the outcome of image categorisation, it is frequently necessary to transform the DN in a satellite image back to TOA reflectance.

5.4.2 Post-processing of Satellite Image

The post-processing of satellite images is done depending upon the user's requirement. User's requirements may like image enhancement, segmentation, fusion, change detection, feature extraction, feature detection, image compression, image classification, etc. Sometimes, visualisation of the feature in the image is the major objective; in such cases, the image enhancement techniques listed below are applied (Asokan et al., 2020):

    i. Filtering with morphological operators
    ii. Histogram equalisation
    iii. Noise removal using a Wiener filter
    iv. Linear contrast adjustment
    v. Median filtering
    vi. Unsharp mask filtering
    vii. Contrast-Limited Adaptive Histogram Equalisation (CLAHE)
    viii. Decorrelation stretch

Any suitable satellite image enhancement algorithm can be used for enhancing the satellite image quality to improve the visibility of features in the image. Figure 5.3 shows the unenhanced satellite image, and Figure 5.4 shows the enhanced satellite image.

Users may be interested in extracting certain specific features like water bodies, forest cover, built-up area, clouds, roads, rivers, canals, etc. from the satellite image. In all such cases, the feature extraction methods given below are utilised (Sedaghat & Mohammadi, 2018; Pare et al., 2018; Rathore et al., 2016):

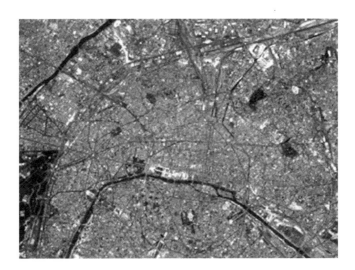

FIGURE 5.4 Enhanced true colour composite

    i. Uniform Competency Feature Extraction
    ii. RepTree, Machine Learning, and Euclidean distance
    iii. Multi-image Saliency Analysis
    iv. Digital Surface Models
    v. Reversible Jump Markov Chain Monte Carlo Sampler

If the user requirement is image segmentation, where the image is divided into segments to locate the objects and boundaries (Grinias et al., 2016), the popular methods of image segmentation are listed below (Suresh & Lal, 2016 and Asokan et al., 2020):

    i. McCulloch's Method, Cuckoo Search
    ii. Deep Convolutional Neural Network
    iii. Markov Random Filed Method
    iv. Gabor Filter, Graph-based Segmentation
    v. In image fusion, two or more images are combined to form a single new image, which is called image merging. With the advent of multiresolution analysis, it is now possible to extract high-frequency components from a multispectral image and subsequently include them into the image as a whole. The wavelet transform is a method that is often used in the process of pan-sharpening. Frequency filtering may be accomplished in a variety of ways, including the High Pass Filter Additive (HPFA) Method, the High–Frequency–Addition (HFA) Method, and the High-Frequency Modulation (HFM) Method, Local Mean Matching (LMM), Local Mean and Variance Matching (LMVM), Regression Variable Substitution (RVS), and Local Correlation Modelling (LCM) are some of the statistical methodologies used to fuse images. These are some of the most often used photo fusion techniques, and they are given below. This program takes

advantage of image cartoon texture decomposition as well as sparse coding techniques:
vi. Contourlet Transform
vii. Expectation Maximisation Algorithm
viii. Dictionary Learning Method

If a user wants to know the change detection in an area after a certain natural event or disaster, the methods listed below could be utilised:

i. Tasselled Cap Transformation (TCT)
ii. Image Differencing
iii. Change Vector Analysis (CVA)
iv. Principal Component Analysis (PCA)
v. Multi-step Image Matching (MSIM)
vi. Breaks for Additive Seasonal and Trend (BFAST)

In a satellite image, if a user is interested in detecting the certain features like cloud, edge, settlement, shadow, etc., methods given below could be utilised:

i. Fuzzy logic-based detection for edge detection
ii. Deep Convolutional Neural Network for informal settlement detection
iii. Invariant pixel detection, PCA for cloud detection
iv. Active contour model for shadows and height detection

There are a plethora of goals that can be achieved with satellite image categorisation, including spatial-temporal data mining and extraction of information for applications, thematic map construction, efficient decision-making, crisis management, visual and digital satellite image interpretation, and field surveys, to name a few. The image classification process will identify and extract the classes of features that we will be generating during the classification process over the course of the classification process. Image classification may be performed under supervision or unsupervised, depending on the circumstances (Abburu & Golla, 2015).

Supervised methods for image classification are as follows: (i) Artificial Neural Network, (ii) Maximum Likelihood, (iii) Bayesian Inference Minimum Travelling Distance, (iv) Parallelepiped, (v) Mahalanobis Distance, (vi) K-Nearest Neighbour, (vii) Decision Tree, (viii) Image Segmentation.

In the case of supervised classification, the training sample of various classes is created from the satellite image. Then various above-mentioned methods are used to classify the image. Whereas in the case of the unsupervised method for image classification, there is no need for a training sample. The algorithm automatically identifies the different classes based on the range of pixel values within the image.

Unsupervised methods for image classification:

i. ISO Data
ii. Support Vector Machine
iii. K-Means

It is clear from the above discussion on satellite image processing that in order to get useful information from the satellite image, a good amount of processing is carried out at various levels which requires high computational power. The reason for high computation is the large number of processes and the large memory size of the satellite image. Many times a single image of the satellite is more than 1 GB. The memory size of a satellite image depends upon the spatial resolution, spectral resolution, and radiometric resolution. Generally, the finer the resolution larger the memory size to store it. Now it is very clear that satellite images have high computational efforts to get useful information from them. On traditional computers, it takes time to process the satellite image. So, processing satellite images on a quantum computer could really be very much time saving. The combination of special-purpose information processors and powerful computers capable of implementing deep neural networks with billions of weights special-purpose information processors and powerful computers have applications in very large datasets and are capable of finding the subtle pattern in data. Quantum computers will be very useful in implementing Machine Learning (ML) and Artificial Intelligence (AI) on big data like satellite datasets.

5.4.3 QUANTUM MACHINE LEARNING

Because of developments in algorithmic breakthroughs and computer capacity, it is now feasible to design complex machine learning algorithms for discovering patterns in data. A wide variety of firms presently adopt these methods. It seems feasible to forecast that in the not-too-distant future, quantum computers will beat conventional computers in machine learning tasks if they can be created in a reasonable period of time. Atypical patterns that are impossible to duplicate under normal conditions are revealed in quantum systems, as opposed to conventional systems (Biamonte et al., 2017). Quantum software is being created and tested in the area of quasi-quantum machine learning, which will allow machine learning to occur at rates far beyond those of normal computers. It has been argued by some researchers that quantum machine learning is a subset of quantum computing as a whole. Even though quantum algorithms have recently been found, there are still major challenges to overcome in terms of both hardware and software before they can be employed as machine learning building blocks. According to current estimates, 2.5 exabytes of data are created every day. There are more than 3.2 billion internet users submitting data to data banks every minute, which is the equivalent of 250,000 Congress Libraries in storage capacity. Additionally, the database currently contains 347,222 tweets and 4.2 million Facebook likes, as well as 9,722 Pinterest Pins, and many more activities, such as downloading documents, taking photographs and videos, and creating accounts. Currently, the database is the world's most comprehensive (Khan & Robles-Kelly, 2020). T.M. Khan and his colleagues were part of the investigation (Khan & Robles-Kelly, 2020). Big data may be utilised to better comprehend complicated biological processes and the causes of climate change, as well as the economic affairs that take place. Climate change may also be researched by looking at the factors that contribute to it. According to current estimates, data generation is rising at a pace ten times faster than computer technology improvements in this field can keep up with. In their paper, Havlíček et al. (2019) distinguished between virtual machine learning, i.e. Qt Modeling Language (QML) and Quantum Applied Machine Learning (QAML). They also distinguished between two types of QML: Quantum Augmented Machine Learning and Quantum Applied Machine Learning (Havlíček et al., 2019). The theories of quantum-inspired machine learning and quantum-generalised learning, which are both consistent with the QML paradigm, also stand out as two very promising research avenues that ought to be explored in more depth. It follows as a result of this that quantum-inspired machine learning cannot achieve the speedups that can be realised with quantum computers; hence, quantum-inspired machine learning is often designed to operate on conventional machines rather than quantum computers. Techniques for quantitative machine learning may be broadly classified in the same manner as conventional machine learning algorithms are: supervised learning, semi-supervised learning, unsupervised teaching/learning, reinforcement learning, and reinforcement learning, to name a few classifications.

Linear support vector machines and perceptrons are two of the most basic supervised machine learning methods currently available. Attempts are made to establish an ideal separation hyperplane between two classes of data in a dataset with the goal of locating all training samples of one class that are either directly or indirectly on one side of the hyperplane with a high likelihood of being on that side of the hyperplane (Vapnik, 1995). In order to improve our chances of developing the most robust classifier for the data, we need to increase the distance between our hyperplane and the data. All of the attributes of the hyperplane are determined by the "weights" that have been learned during the training phase. Along with its ability to accommodate non-linear hypersurfaces via the use of kernel functions, the support vector machine has a number of other advantageous characteristics. Classifiers of this kind have shown to be highly useful in a variety of applications, including picture segmentation and the biological sciences. Grover's search for function minimisation is the basis of a quantum support vector machine (QSVM), which has been proposed since the early twenty-first century and is based on a variation of Grover's search for function minimisation. This happened during the first decade of the twenty-first century (Anguita et al., 2003).

The quantum state of an n-bit quantum bit or qubit may be represented as a vector in a 2^n-dimensional complex vector space, which is called a complex vector space. By multiplying a pair of $2^n \times 2^n$ matrices by a quantum logic operation or a quantum measurement on a qubit, we may get the corresponding state vector, which is a basic aspect of the techniques discussed in the next section. In recent years, several researchers have discovered that the construction of such matrix transformations enables quantum computers to perform linear algebraic operations such as Fourier transforms (Shor, 1997), the determination of eigenvectors and eigenvalues (Nielsen & Chuang, 2000), and solving linear sets of equations over $2n$-dimensional vector spaces in time that is polynomial in n, which is exponentially faster than their most well-known classical counterparts (Shor et al., 1994). Several names have been used to refer to the algorithm invented by Harrow, Hassidim, and Lloyd (sometimes referred to as the HHL algorithm), which is used for linear systems of equations (Harrow et al., 2009). Consider the following example: It may be required to iterate over a pool of support vectors N times in order to identify s support vectors from a pool of N support vectors. Most recently, researchers developed a least-squares quantum support vector machine that makes extensive use of the quantum Basic Linear Algebra Subroutines (qBLAS) that were discovered throughout the course of their inquiry (Childs et al., 2015). Several types of data sources, including a quantum subroutine that generates quantum states and a quantum Random Access Memory (qRAM) that reads conventional data, may be used to generate the necessary input data. Quantum phase estimation and matrix inversion methods are applied to the data as soon as it is made accessible to the quantum computing equipment, allowing the necessary results to be obtained as quickly as possible (Biamonte et al., 2017). This paragraph demonstrates that, theoretically, it is feasible to construct the ideal separation hyperplane and determine whether a vector is on one side or the other of the hyperplane in less time than it takes to read this paragraph.

With the aid of quantum annealers, which are easily accessible on the market and can be found in a variety of configurations, deep quantum learners may be deployed in a practical setting. The D-Wave quantum annealer, which is a transverse Ising model that can be used in a single step to construct the thermal states of both classical and quantum spin systems, has the capability of creating the thermal states of both classical and quantum spin systems. It has a programmable interface and may be used for a number of tasks. This device has been shown to be capable of running deep quantum learning algorithms on thousands of spins, and the findings have been published in a peer-reviewed publication as evidence of this capability (Benedetti et al., 2016). A new generation of quantum Boltzmann machines with more flexible couplings is currently being developed in order to improve upon the existing technology in order to enable universal quantum logic. In theory, deep quantum networks may learn how to create quantum states that are characteristic of a broad variety of systems over time, but this does not seem to be the case at this point. Future deep quantum networks might possibly perform functions comparable to those of a quantum associative memory for a wide range of various sorts of systems. Classical machine learning approaches, on the other hand, are incapable of generating quantum states and must be abandoned. Quantum Boltzmann training, as shown in the following examples, may be used for a variety of applications, including machine learning, in addition to the development of more accurate models for classical data and the classification of quantum states. In order for a quantum computer to be used to assist in the characterisation of a quantum system or to accept input states for use in a quantum PCA approach, the enormous technical challenge of loading coherent input states into a quantum computer must be overcome before the technique can be implemented. These applications will continue to rank among the most promising applications for quantum machine learning in the foreseeable future because they do not rely on quantum RAM and may be able to achieve exponential speedups in device characterisation in the near future.

5.5 CONCLUSION AND FUTURE SCOPE

Quantum computing can aid in the more accurate and timely weather forecast, reduce the simulation and testing time for satellite component which requires rigorous testing, the selection of molecules for the production of organic batteries, and drug modelling and testing, among other specific scientific challenges. The quantum key distribution technique will be the most advanced form of secure communication in the near future. Method security in communication will be rendered obsolete once quantum computers are fully operational, since quantum computers will be able to simply crack the security code used to protect the communication. As a result, quantum key distribution will be the answer to the problem of secure communication. There are many technical challenges that must be resolved, the most important of which is how to distribute the quantum key across great distances. It is necessary to conduct additional experiments in order to tackle these scientific challenges. Because the existing technique for distributing quantum keys relies on photon transmission, photon losses occur when the key is transmitted over a long distance. As a result, there is far more noise than signal. Accurate optical qualities of the atmosphere are required in order to reduce noise levels. In the event that we are successful in learning how to communicate information through quantum entanglement, this will be a game-changing discovery. Classical machine learning can also be used to derive theoretical insights into quantum states. The use of neural networks to examine phase-of-matter detection and ground-state search has lately become popular in condensed matter research. These were able to outperform more well-known mathematical tools. Theoretical physicists are currently looking at these models to see how they compare to classic methods like tensor networks in terms of their descriptive capability. Exotic state of matter applications is already on the market, and they have been shown to capture extremely non-trivial aspects from disordered or topologically ordered systems.

The use of learning methodologies for the development of control sequences that optimise adaptive quantum metrology has shown to be quite beneficial. Adaptive quantum metrology is a critical component of many quantum technologies, and it is discussed in detail below. Evolutionary algorithms for quantum molecule control have been developed in order to solve the difficulties presented by changing ambient factors during an experiment. Techniques for rewarding learning that make use of heuristic global optimisation, such as those employed in circuit design, have shown to be quite effective, particularly when it comes to noise and decoherence. They have also shown that they scale effectively with the size of the system. Reinforcement learning has additional advantages over traditional quantum systems, such as gate-based quantum systems and regular quantum systems. Intelligent agents, when paired with quantum information to provide adaptive calibration, are used in adaptive controllers to correct for an external stray field with uncertain amplitude and direction when the amplitude and direction are unknown.

The potential for productivity and scalability of quantum computing is enormous when compared to standard computing methods such as computers and servers. Whether or not this promise will be fulfilled in its entirety in practise, however, remains to be seen. As a matter of fact, it is commonly believed that every issue that a quantum computing paradigm may be able to solve may also be addressed by using a conventional computer. As a consequence, since quantum computers are predicted to attain efficiencies that need far lower quantum integration requirements for equivalent computing workloads than conventional machines, a huge scale of integration would be required. Moreover, when applying quantum computing to data coming from non-quantum settings, which are widespread in consumer applications and computer research but not in quantum phenomena themselves, there are additional considerations to be made.

A lot of obstacles must be overcome on both the hardware and software sides of the equation before quantum machine learning can be properly implemented. First and foremost, it is necessary to ensure that quantum hardware is technically feasible in order to reap the benefits of quantum algorithms. A second time, the integration of interface devices such as qRAM is required in order to encode conventional data into a quantum mechanical form using Quantum Mechanical Logic. There is no denying that these hardware flaws are severe, and they must be addressed immediately. Finally, in order to fully implement Quantum Mechanical Logic approaches, it is required to take into account the constraints of the application of quantum algorithms in general.

There are four major issues with quantum computing, which are discussed as follows:

- It is essential to know an exponential number of bits in order to get the whole result as a string of bits from different quantum algorithms, which is a difficult obstacle to overcome. As a result, a number of Quantum Mechanical Logic approaches have been rendered obsolete. It is also possible that this problem may be avoided if just summary data about the solution state were made accessible.
- The issue with the input is as follows: In spite of the fact that quantum algorithms have the potential to significantly speed data processing, they have so far failed to produce substantial advances in data reading. Thus, in a limited number of cases, the cost of reading input data may be more than the cost of performing quantum algorithm operations itself, as seen above. Learning all there is to know about this component is an ongoing process.
- When it comes to benchmarking, the following approach is taken: For the simple reason that there is extensive benchmarking that has to be done against present heuristic techniques, claiming that a quantum algorithm is always better than all known conventional machine algorithms is difficult. A reduction in the Quantum Mechanical Logic restrictions would go a long way toward partially resolving this problem.
- Due to the price issue, which is closely tied to the input/output difficulties outlined above, it is still questionable how much money will be spent on quantum machine learning algorithms in the near future. Complexity constraints show that they will provide considerable benefits for sufficiently big situations if they are applied correctly.

We should use quantum computing on quantum data instead of classical data to circumvent some of these issues. One goal is to characterise and operate quantum computers via quantum machine learning. This will start a virtuous cycle of the invention similar to that seen in traditional computing, in which the previous generation of processors is used to build the next generation of processors. The first fruits of this cycle have already begun to appear, as classical machine learning enhances the quantum processor which will in return lead to an exponential increase in the computational power through quantum-enhanced machine learning.

REFERENCES

Abburu S. & Golla, B.S. Satellite image classification methods and techniques: A review, *Int. J. Comput. Appl.*, 119, 20–25 (2015).

Anguita, D., Ridella, S., Rivieccio, F. & Zunino, R. Quantum optimization for training support vector machines. *Neural Netw.*, 16, 763–770 (2003).

Aoki T. & Shimokawabe T. Large-scale numerical weather prediction on GPU supercomputer. In: Yuen D., Wang L., Chi X., Johnsson L., Ge W. & Shi Y. (eds) *GPU Solutions to Multiscale Problems in Science and Engineering. Lecture Notes in Earth System Sciences.* Springer, Berlin, Heidelberg (2013). https://doi.org/10.1007/978-3-642-16405-7_16.

Asokan, A., Anitha, J., Ciobanu, M., Gabor, A., Naaji, A. & Hemanth, D.J. Image processing techniques for analysis of satellite images for historical maps classification—an overview. *Appl. Sci.*, 10(12), 4207 (2020) https://doi.org/10.3390/app10124207.

Bacsardi, L. On the way to quantumbased satellite communication. *IEEE Commun. Mag.*, 51(8), 50–55 (2013).

Biamonte, J., Wittek, P., Pancotti, N. et al. Quantum machine learning. *Nature*, 549, 195–202 (2017).

Burrows, W.E., *This New Ocean: The Story of the First Space Age*, The Modern Library, New York (1999).

Childs, A. M., Kothari, R. & Somma, R. D., Quantum linear systems algorithm with exponentially improved dependence on precision (2015). Preprint at https://arxiv.org/abs/1511.02306.

Clerk, A.A., Lehnert, K.W., Bertet, P. et al. Hybrid quantum systems with circuit quantum electrodynamics. *Nat. Phys.*, 16, 257–267 (2020). https://doi.org/10.1038/s41567-020-0797-9.

Grinias, I., Panagiotakis, C. & Tziritas, G. MRF-based segmentation and unsupervised classification for building and road detection in peri-urban areas of high-resolution satellite images. *ISPRS J. Photogramm. Remote Sens.*, 122, 145–166 (2016).

Harrow, A. W., Hassidim, A. & Lloyd, S. Quantum algorithm for linear systems of equations. *Phys. Rev. Lett.*, 103, 150502 (2009).

Havlíček, V., Córcoles, A.D., Temme, K. et al. Supervised learning with quantum-enhanced feature spaces. *Nature*, 567, 209–212 (2019).

Khan T. M. & Robles-Kelly A., Machine learning: Quantum vs classical. *IEEE Access*, 8, 219275–219294 (2020).

Liao, SK., Cai, W.Q., Liu, WY. et al. Satellite-to-ground quantum key distribution. *Nature*, 549, 43–47 (2017). https://doi.org/10.1038/nature23655.

Lillesand T.M., *Remote Sensing and Image Interpretation*, 6th Edition, John Wiley & Sons, New York (2016).

Benedetti, M., Realpe-Gómez, J., Biswas, R. & Perdomo-Ortiz, A. *Phys. Rev.* A 94, 022308 (2016).

Mastriani, M., Iyengar, S.S. & Kumar, L. Satellite quantum communication protocol regardless of the weather. *Opt. Quant. Electron.*, 53, 181 (2021).

Matuschek, O., & Matzarakis, A. A mapping tool for climatological applications. *Meteorol. Appl.*, 18(2), 230–237 (2010). https://doi.org/10.1002/met.233.

Nielsen, M. A. & Chuang, I. L. *Quantum Computation and Quantum Information*, Cambridge University Press (2000).

Villoresi, P., Jennewein, T., Tamburini, F. et al. Experimental verification of the feasibility of a quantum channel between space and earth, *New J. Phys.*, 10(3), 033038 (2008). https://doi.org/10.1088/1367–2630/10/3/033038.

Pare, S., Bhandari, A.K., Kumar, A., & Singh, G.K. A new technique for multilevel color image thresholding based on modified fuzzy entropy and Lévy flight firefly algorithm. *Comput. Electr. Eng.*, 70, 476–495 (2018).

Shor, P.W. In Proceedings of the 35th Annual Symposium on Foundations of Computer Science (*IEEE Computer Society Press*, 1994), pp. 124–134.

Rathore, M.M.U., Ahmad, A., Paul, A. & Wu, J. Real-time continuous feature extraction in large size satellite images. *J. Syst. Archit.*, 64, 122–132 (2016).

Rivest, R., Shamir, A. & Adleman, L. A method for obtaining digital signatures and public key cryptosystems, *Commun. ACM*, 21(2), 120–126 (1978).

Sedaghat, A. & Mohammadi, N. Uniform competency-based local feature extraction for remote sensing images, *ISPRS J. Photogram. Remote Sens.*, 135, 142–157, (2018).

Shor, P. W. Polynomial-time algorithms for prime factorization and discrete logarithms on a quantum computer, *SIAM J. Comput.*, 26, 1484–1509 (1997).

Smith, G. Newton's philosophiae naturalis principia mathematica. In: Zalta E. N. (ed.) *The Stanford Encyclopedia of Philosophy*. Winter edition (2008). http://plato.stanford.edu/archives/win2008/entries/newton-principia/.

Suresh, S. & Lal, S. An efficient cuckoo search algorithm based multilevel thresholding for segmentation of satellite images using different objective functions. *Expert Syst. Appl.*, 58, 184–209 (2016).

Vapnik, V. *The Nature of Statistical Learning Theory*, Springer, New York (1995).

Villoresi, P., et al.: Experimental verification of the feasibility of a quantum channel between space and Earth. *New J. Phys.* 10 (2008).

Zhang, J., Qiu, X., Li, X., Huang, Z., Wu, M. & Dong, Y., Support vector machine weather prediction technology based on the improved quantum optimization algorithm. *Comput. Intell. Neurosci.*, 2021, 6653659 (2021).

6 Evolution of Deep Quantum Learning Models Based on Comprehensive Survey on Effective Malware Identification and Analysis

S. Poornima and Thiruselvan Subramanian
Presidency University

CONTENTS

DOI: 10.1201/9781003250357-6

6.1 INTRODUCTION

Nowadays, security threat is the biggest issue faced by IT, private, and government organizations, since they need more amount of information has to be transferred end to end every minute. Hackers are the ones, who adds malware's to the original content during transferring which leads to content demolish, content modification, content leaks, and also to violation of Intellectual Property Rights (IPR). Therefore, there is in need of effective malware detection and analysis at nodes to avoid the security threats.

6.1.1 CLASSICAL MACHINE LEARNING

In the early 1950s, AI emerged to prove that machines can perform a task equally to humans. AI performed very well but produced less accuracy. ML was derived from AI to improve accuracy by adopting the probability and statistics theory (Wang et al., 2018). A famous scientist, Arthur Samuel, researched ML and came up with his popular definition in 1959, as ML provides computer systems with the ability to learn without explicitly being programmed (Faruki et al., 2012). After many scientists raised their opposition such that the ML algorithms are not adapting the learning by themselves, and it is an external method that works when it is called. More formal languages define the software with respective Input-Output integration works upon training data already derived by users. These types of software enhance human-computer interaction and provide flexibility in updating user needs quickly. The email spam filtering technique is considered the best-suited example, which learns new arising data by observing user behavior and utilizes a data repository to classify new spam emails. Machine learning is integrated with human lives by offering many user services day to day.

6.1.2 COMPARISON BETWEEN MACHINE LEARNING AND DEEP LEARNING

ML and DL domains are derived from AI. At the same time, DL is considered a part of ML. However, DL differs in terms of adopting conventional ML techniques. ML offers more user benefits in predicting real-time scenarios, but with few drawbacks like variances, model complexity, etc. DL overcomes the disadvantages of ML by analyzing more feature sets from data. ML improves system performance when the dataset is huge. But, DL consists of layer performances to analyze the dataset. Table 6.1 shows the differentiation between ML and DL with various parameters.

ML was invented as the integration of two domains—AI and Statistics. ML systems perform very well when a large dataset is used for training, improving accuracy whenever a new problem statement is inputted. The more significant advantage of an ML system is that it is not dependent on a computer system or hardware used as it can be deployed on any of the systems without any conflicts. ML algorithms reduce data complexity by necessary encoding procedures, and ease data analysis procedures, leading to improved accurate solutions. The algorithms initiate the process by feature

TABLE 6.1

Differentiating Machine Learning and Deep Learning Models with Parameters

Parameter	Machine Learning	Deep Learning
Data dependency	It works on a smaller amount of data, even though a huge amount is dataset is fed.	Deep learning algorithms rely on a huge data repository for increased output accuracy.
Runtime	It consumes less time to train the model than deep learning.	It requires significant runtime to train and test the model, mainly General Processing Unit (GPU) needs to be purchased.
Hardware dependency	They work on general Central Processing Unit (CPU) configuration since it needs fewer data to be computed.	To process huge data repositories, it needs GPUs and high-end machines.
Feature engineering	It is needed and performed by the expert to process the model.	It does not require feature engineering, but it tries to learn high-level features from the data on its own.
Problem-solving approach	It divides the dataset into partitions, performs classification linearly, and represents the final predictions.	It processes the dataset as a whole and produces the final result instantly.
Result interpretation	Interpreting the result for a given problem is easy by using various validation metrics.	Interpreting the result for a given problem is complex because the network acts as a black box, and the programmer cannot understand the processing method used.
Data type	It requires structural data representation.	It processes all types of data like structured and unstructured data for problem-solving.
Applications	It is used to solve both simple and complex problems.	It is used to solve complicated problem statements according to user requirements.

engineering, then perform cross-validation on the dataset, then implement various classifiers and cumulate their outputs. It utilizes the divide and conquer method for extracting outcomes from data analysis.

Artificial Neural Network (ANN) is alternatively called deep learning (DL), derived from ML and AI. DL is used to implement neural networks according to the data sample inputted. It produces more accurate outputs than ML since it allows more time for training the network. Clusters are created based on similar inputs; so, it will be easier to retrieve the outputs from all the clusters created leading to improved solutions. DL deals with a huge quantity of datasets for effective data analysis.

Therefore advanced computers are required for its execution. Many real-time applications utilize DL techniques to offer necessary flexible user services, like image recognition systems, voice recognition systems, biometric applications, bioinformatics, etc. Simultaneously, it can be integrated into medical imaging applications and can be formatted via parameters to get desired solutions. DL can be assimilated with neural networks to form various network architectures like Artificial Neural Networks (ANNs), Deep Neural Networks (DNNs), and Convolutional Neural Networks (CNNs). These network architectures function based on the working principle of the human brain. It solves real-time problems by just exaggerating the working nature of the human brain with improved functionalities.

In recent years, DL and ML have played a major role in market trends. Therefore, each software firm demands the data scientist to have expertise in ML and DL. The above-mentioned models perform the problem-solving cost-effectively since the models are implemented in multi-disciplinary problem statements. More data researchers are arising and trying to research the above learning techniques to the core to benefit the users (Nataraj et al., 2011). Definitely, these learning techniques will be utilized in many newer real-time applications with improved accurate solutions according to the problem statements. Data researchers found machine learning models as it will provide a bright future in the neuroscience field.

6.1.3 RISE OF QUANTUM COMPUTING IN MACHINE LEARNING

The ideas from quantum computing (QC) and ML motivated the invention of QML being considered a significant research area nowadays. QML helps the users add multiple hardware as a new computing machine named quantum computer (Mohaisen et al., 2015). Consequently, it assisted ineffective data manipulations on quantum machines based on quantum theory, a varied law of the physics domain. It leads and entangles quantum computing differing in the process of manipulation activities (Dinaburg et al., 2008). It is found that quantum computers will have control over data, unlike traditional computers. Figure 6.1 explains the overview of the ML system replaced with QML procedures which leads to increased predictive output.

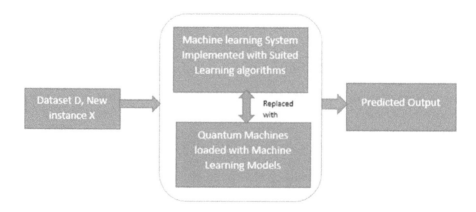

FIGURE 6.1   Emergence of classical machine learning to quantum machine learning

A quantum computer employs qubits called quantum bits; it works on Boolean logic and toggles its output between 0 and 1. The quantum computer's output is validated by considering the qubits count along with its associated functions, which satisfies the user's basic needs. QML utilizes quantum computers to adapt suitable learning algorithms for the defined problem statements. QML computers, which require the integration of qubits and logic gates, are called variational algorithms, and they can be executed on certain quantum machines only. The learning algorithm varies based on the user feasibility and its architecture. Many algorithms like Grover's algorithm and Shor's algorithm are employed based on a varying number of qubits (Hu et al., 2013). Still, now, few benchmark standards are derived from executing ML algorithms on real quantum architectures.

Efficient information processing is the major motivation lying behind QC. Analysis of quantum states leads to fast computations on ML models regarding computational complexities like time and storage because QC benefits the user in executing multiple operations instantly. Qubit is represented as $|\psi i = \alpha |0i + \beta |1i$ (with α, $\beta \in C$ and $|0i$, $|1i$ in the two-dimensional Hilbert space H2). The quantum states are validated based on logic 0 or logic 1, and it is maintained throughout the application execution supported by probability conservation as $|\alpha|^2 + |\beta|^2 = 1$. In probability and statistics, the change of one quantum state to another is said to be unitary. QC provides distinct qubit design gates leading to manipulation of basic parameters like amplitude and its phase. For example, the distinct qubit design will have 3D gates visibility namely X, Y, and Z gates. In this particular design, the qubit will be assigned with $\beta = 0$ ($\alpha = 0$) during its processing. Table 6.2 differentiates QC from classical computing and the motivation to adopt it.

6.1.4 EVALUATION METRICS

ML and DL models were validated to determine the performance measure of the system via statistical evaluation metrics. For any real-time application, it is mandatory

TABLE 6.2
Differentiating Classical Computing and Quantum Computing

Validation Parameters	Classical Computing	Quantum Computing
Processing bit	Binary 0 and Binary 1 are used for computations.	Computations are performed using qubits, which can represent 0 and 1 in parallel.
Power consumption	1:1 transistor integration is utilized to compute the power consumption.	Based on the utilization of qubit counts, power consumption is computed.
Cost	Classical computers have improved accuracy and can be employed at room temperature.	Quantum computers have reduced accuracy and need to be kept ultra-cold.
System performance	Traditional Computers results in very low processing leads to increased time complexity.	Complicated tasks can be employed.

to evaluate the models imposed in it. Many evaluation metrics exist to test a model, like confusion matrix, logarithmic loss, accuracy, residual analysis, logarithmic loss, etc. All the validation parameters can be combined together to estimate the performance of the ML/DL model.

Table 6.3 represents the evaluation metrics to validate the ML and DL models used. During the classification task, the confusion matrix is the adaptable metric to best validate the binary classification problems, as mentioned in Table 6.3. In Table 6.3, actual classes are represented in a column manner and predicted classes are represented in a row manner. While computing the confusion matrix, *tp* and *tn* indicate the positive and negative instance counts, respectively, that are classified correctly, and *fp* and *fn* indicate the positive and negative instance counts, respectively, that are misclassified. Table 6.2 represented the validation metrics used to evaluate the ML and DL models' performance with different viewpoints on evaluation procedures. But still, multi-class problems lead to reduced performance; therefore, a few more metrics were represented in Table 6.3 for problems with multi-class classification.

6.2   MALWARE DETECTION AND ANALYSIS (MDA)

In this Internet era, malware is the primary threat in society and increasing in numbers and attacks. Several anti-malware techniques and software are invented, even though there are still unknown malware to be addressed. This chapter discusses the detection and analysis of malware integrated into the dataset and determines the type of identified malware and its attacks. The malicious data is incorporated to steal, damage, and corrupt the target user's vital information. Nowadays, the malware inventors spread their malicious data via malware software like spyware, viruses, and executables and now initiated to incorporate malicious code in open-source available datasets. The open-source software users unknowingly download the malicious dataset and utilize it for problem-solving, resulting in their data loss. Antivirus developers still detecting increasing malware day to day and also with higher impact. In such prevailing situations, just firewalls and antivirus scanning is just not enough to detect and vanish malware.

Nowadays, there exist freely downloadable codes in all languages on many Internet websites. Malicious codes are generated by attackers with the motivation of their noxious code scattered over the society. This includes the following:

- Thieving trustworthy data from others for their economic gain, e.g., selling users' personal information like credit cards, bank details, etc.
- Damaging servers or systems by creating false alarms leading programmers to deal with technical issues
- To destroy opponent's business, by stealing their confidential data
- Stealing government digital assets and doing terroristic attack leading to economic loss and human lives

This work presents the list of malware types, tools, and techniques used for effective malware detection and analysis. It also focuses on open-source Malware Detection

TABLE 6.3

Evaluation Metrics used for ML/DL Models

Metrics	Formula	Evaluation Forms
Accuracy	$\dfrac{tp+tn}{tp+fp+tn+fn}$	Measures the ratio of correct prediction over the total number of instances evaluated
Error rate	$\dfrac{fp+fn}{tp+fp+tn+fn}$	Measures the ratio of incorrect predictions over the total number of instances evaluated
Sensitivity	$\dfrac{tp}{tp+fn}$	Measures the function of positive instances that are correctly classified
Specificity	$\dfrac{tn}{tn+fn}$	Measures the function of negative patterns that are correctly classified
Precision	$\dfrac{tp}{tp+fp}$	Measures the positive instances that are correctly predicted from the total predicted patterns in a predictive class
Recall	$\dfrac{tp}{tp+tn}$	Measures the fraction of positive patterns that are correctly classified
F-measure	$\dfrac{z+p+r}{p+r}$	Represents the harmonic mean between recall and precision values
Geometric mean	$\sqrt{tp+tn}$	Maximizes the tp rate and tn rate and simultaneously keeps both rates relatively balanced
Averaged accuracy	$\dfrac{\sum_{i=1}^{n} \dfrac{tp_i + tn_i}{tp_i + fn_i + fp_i}}{t}$	Average effectiveness of all classes
Averaged error rate	$\dfrac{\sum_{i=1}^{n} \dfrac{fp_i + fn_i}{tp_i + fn_i + fp_i}}{t}$	Average error rate of all classes
Averaged precision	$\dfrac{\sum_{i=1}^{n} \dfrac{tp_i}{tp_i + fp_i}}{t}$	Average of per-class prediction
Averaged recall	$\dfrac{\sum_{i=1}^{n} \dfrac{tp_i}{tp_i + fn_i}}{t}$	Average of per-class recall
Averaged F-measure	$\dfrac{2+P_M+T_M}{P_M+T_M}$	Average of per-class F-Measure

and Analysis (MDA) tools that rely on cloud computing to provide a complete malware analysis report. Moreover, a list of algorithms used for malware analysis were focused and also malware analysis tools were suggested for appropriate problem statements.

Many cybersecurity marketers offer malware detection and analysis as a service to corporate users. This helps the corporate people to detect whether the downloaded file is hided with malware to prevent the system from its attacks and stop it from spreading. Due to this, more security mechanisms were identified and made public the thwart future attacks. Figure 6.2 represents the growing graph of malware counts in society and shows that the malware increased from 3.0 USD billion to 11.7 USD billion from 2019 to 2024, and it is captured by 31.0 Compound Annual Growth Rate (CAGR) during the same time period (Abusnaina, 2021).

Existing key market companies focus on producing malware analysis reports with their effects. These IT concerns also offer solutions by analyzing the market's current needs. The malware analysis got popularized by the IT firms by partnering generating Memorandum of Agreements with other firms so that they can launch new products along with their own versions.

The Statista Survey produced a malware report in March 2020, stating that new malware were upcoming and detected daily in 677.77 million software in the end of January 2020. AV-TEST also stated that 700 million new malware will be identified within 2020 (Figure 6.3).

This paper presents the following contributions:

- **Existing Malware Detection and Analysis Algorithms:** This paper focuses on explaining existing malware detection and analysis algorithm and comparing them with various domains and their severity in attacks. Also, this section defines the tools and techniques existing in the market and their usage.

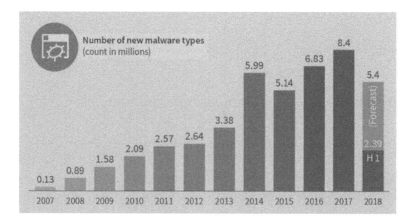

FIGURE 6.2   Market survey analysis on malwares (Mohaisen et al., 2015)

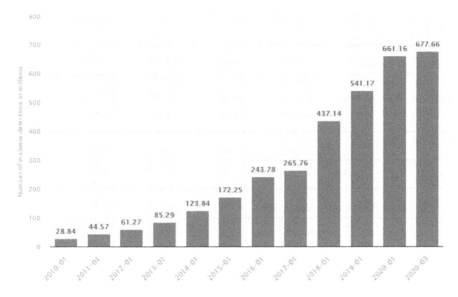

FIGURE 6.3   Survey on types of malware growth (Mohaison, 2020)

- **Survey on Malware Analysis:** The pros and cons of prevailing surveys in this area are discussed along with their reasons. Different viewpoints are considered in pinpointing the study, and reasons for the failure of those surveys are explained.
- **Wide-Ranging Malware Detection Tools:** More corporate vendors released their own version of malware detection and analysis tools in recent years. This paper produces a comprehensive list of available tools regarding malware detection, malware analysis, obfuscators, reverse engineering, etc. Also, more factors like domain, tools, and techniques in this area are explained in detail.

6.3   MALWARE DETECTION—A VIEW

Malware detection is considered vital as malware destroys and steals users' data. It is defined as the technique of detecting malware in downloaded software programs by scanning the complete data and differentiating the original program from malware-affected programs. This detection process enables the user to protect their data, thereby safeguarding IPR. The malware detection process is a variety of malware detection tools and techniques that exist in the market and differ in terms of time and space consumption. This process consumes less time complexity to detect any malware associated with the free downloading programs. Malware is becoming solid and stealthy every day; so, to analyze the newer malware attacks and impacts, it is mandatory to develop new malware detection algorithms up to date. The malware detection process involves various phases that are classified as signature-based detection, heuristic-based detection, and specification-based detection.

6.3.1 SIGNATURE-BASED DETECTION

This method detects virus codes to detect malware. Each malware carries a unique virus code, which helps the detectors to identify the type of virus and its impacts. When a user attempts to receive a file from various sources, a malware-infused file reaches the system. The detector immediately scans the entire file, identifies the code incorporated in it, and sends it to the authorized huge data repository. The respective virus code is searched in the data repository; if it is found, it is recognized as a particular virus; and a report is sent to the host system. Immediately, the detector denies the file from downloading and erases it along with its source files. Suppose the malware detector identifies a new virus code based on its functionality. In that case, it will be categorized as a new malware and added to the data repository to be protected from such malware attacks in the future.

6.3.2 HEURISTIC-BASED DETECTION

This detection is also known as behavior- or anomaly-based detection. This technique is majorly intended for identifying the known and unknown malware's behavior. This detection methodology identifies malware based on malware's origin address, attached formats, statistical symbols, etc. It consists of two stages – training and the detection. The training stage observes the system's behavior during the absence of malware attacks. In the detection stage, the system's varied behavior is captured and compared with the initial behavior; the variances found are labeled as malware attacks. Figure 6.4 shows the behavior-based detection methodology.

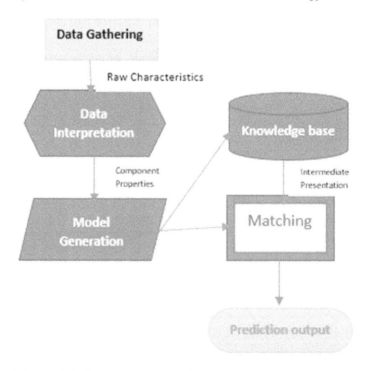

FIGURE 6.4   Heuristic-based malware detection process

6.3.3 Specification-Based Detection

This method strives to derive program specifications from capturing the behavior of crucial software. Apart from its specifications, it also consists of program monitoring events, executional factors, and differences in behavior flags for effective detection. This detection mechanism builds its own specifications for identifying particular malware attacks and helps to protect the data specifically. Flags like False Positive (FP) and False Negative (FN) indicate the system's behavior variances. The major drawback of this method is that it cannot observe the system's entire behavior and cannot classify the varying system behavior. For complex problem statements, this technique is not applicable. Even though, this technique derives its own specifications, but still it is unable to represent the current state of the systems at an instant.

6.4 MALWARE ANALYSIS—A VIEW

The malware analysis technique is used to obtain the malware report, and it consists of the type, parts, and functions of attacked malware. To perform malware analysis, two techniques—static analysis and dynamic analysis—are adopted. The process of malware analysis without executing it is known as static analysis. Static analysis is further divided into two phases—basic static analysis and advanced static analysis. The process of executing the malware and analyzing its impacts is known as dynamic analysis. The static analysis method is safer than dynamic analysis. However, dynamic analysis can be performed under virtual mode. Therefore, it does not destruct the system. Dynamic analysis is again split into two phases—fundamental dynamic analysis and advanced dynamic analysis. It is shown in the above Figure 6.5.

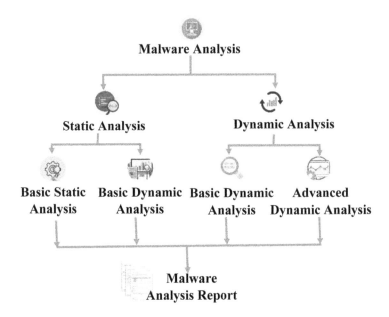

FIGURE 6.5 Categories of the malware analysis process

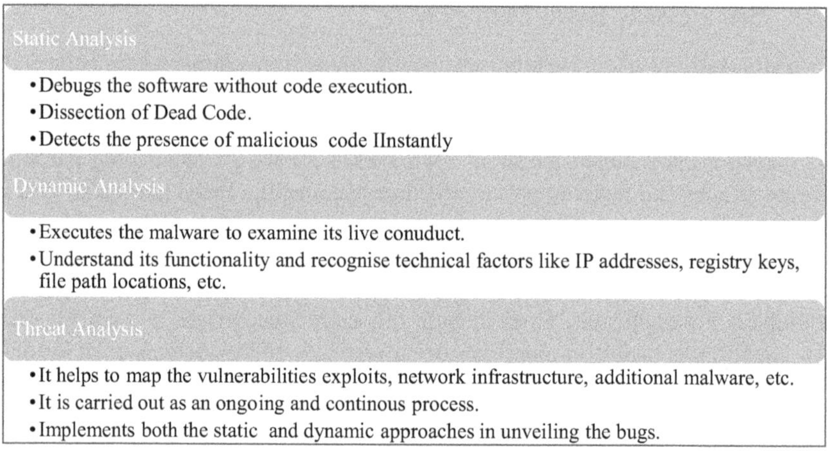

Static Analysis
• Debugs the software without code execution.
• Dissection of Dead Code.
• Detects the presence of malicious code IInstantly

Dynamic Analysis
• Executes the malware to examine its live conuduct.
• Understand its functionality and recognise technical factors like IP addresses, registry keys, file path locations, etc.

Threat Analysis
• It helps to map the vulnerabilities exploits, network infrastructure, additional malware, etc.
• It is carried out as an ongoing and continous process.
• Implements both the static and dynamic approaches in unveiling the bugs.

FIGURE 6.6   Types of malware analysis process (Lee et al., 2011)

For developing an ML model, malware of a similar type are clustered in a group since it exists in various forms. They are divided based on their harness, as they can exist in more classes. Therefore, malware analysis is derived from the detection procedure to identify the complete sustainability of malware. To analyze the malware, its function-alities, effects, and motivation, malware analysis is a much-needed step that helps the programmer design anti-malware software. Based on the tools and algorithms adopted, malware analysis is partitioned into three classes and shown in Figures 6.5 and 6.6.

6.4.1 MALWARE ANALYSIS PROCESS

Malware analysis refers to scanning the malware and generating a report on it via a four-stage process, as shown in Figure 6.7. As the process gets expanded, the compu-tational complexity also increases equally to its functionalities.

1. **Fully Automated Analysis:** It detects malware based on labeled informa-tion stored previously in the data repositories. The advantage of this method is that it is fast enough to generate the analysis report and provides malware ratings based on machine configuration.
2. **Static Analysis:** It is the process of discovering a file's attribute without executing it. This process is most suited for executing files in the virtual mode so that malware may not destruct the system data.
3. **Interactive Behavior Analysis:** In this phase, the file incorporated with malware is executed in virtual mode, and its behavior is captured like its communication with the machine, modifying controls, stealing data, etc. There are a few malware that will not be executed in virtual mode, since it needs interaction with its host system. In such cases, a red flag is raised on permissions like modifying existing data, installing a new process, chang-ing system registry, etc.

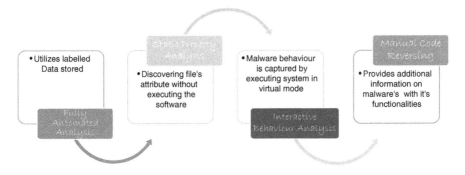

FIGURE 6.7   Malware analysis framework

4. **Manual Code Reversing:** This process provides additional information on malware by reversing the code manually. It is considered an advantage, since the additional information is used to identify the tools and techniques behind the malware, it defines the code functionalities, and identifies the host system from where it is generated.

6.4.2   MALWARE ANALYSIS APPROACH USING MACHINE LEARNING/DEEP LEARNING MODELS

In previous sections, malware detection and analysis were discussed in detail. Various techniques such as instruction embedding, register assigning, code obfuscations, and opcode embedding were also discussed. This section focuses on a comprehensive analysis of multiple datasets implemented via ML and DL techniques. ML and DL employ their own set of algorithms to detect and analyze malwares. They include static analysis and dynamic analysis mechanisms to validate the dataset attributes. By considering the ML and DL technique, this section represents a detailed review of existing ML and DL tools and techniques along with pros and cons. Moreover, report generated from variant behavior, instruction opcodes, Application Programming Interface (API) calls, and opcode validation feature set is considered for legitimating the malicious samples. In summary, the literature review of malware analysis via ML and DL is represented by considering the following parameters.

Malware detection is widespreadly analyzed by employing ML and DL models considering Program Executable (PE) files, ransomware files, android applications, and metamorphic data as input data source. It is shown in previous sections that traditional ML algorithms utilize both theoretical datasets and static feature engineering processes for improved error rate detection. It is found that later DL techniques produce enhanced predictions in all types of problem statements. Existing malware detection systems were evaluated with various classification techniques by considering opcode sequencing and API/system calls as features. The classification model performances are validated based on varying discrete and continuous variables. We found that the integrations of discrete and continuous attributes are not improving model predictions. While comparing DL and ML techniques, it was proven that an improved

F1 score can be employed for exhaustive validation. Additionally, code obfuscations were analyzed with enhanced hyperparameters on samples using existing virus kits.

Therefore, we finalized that system affected with varying malware creates metamorphic effects leading to system crashes too. These malware kits integrate obfuscated codes to identify the project flow of execution and their functions. Here, it is proved that ML techniques can easily detect the obfuscated code via generic features like opcode sequencing, etc. Table 6.4 shows the literature survey based on malware analysis with their implementational models along with their pros and cons.

6.5   DISCUSSION

The previous section provided a review of malware detection and analysis with ML and DL techniques. It is observed that in the collection of malware detection techniques, 65% methods employ static analysis, 15% employ dynamic analysis, and the remaining employ hybrid analysis techniques. From the review studies, static analysis was adopted by most of the ML/DL techniques due to its benefits like vulnerability detection, cost benefits, etc. Drebin is the dataset used in widespread for android malware detection (Wang et al., 2016). The next preferred datasets are Google Pay, MelGenome, and AMD. Drebin provides a labeled dataset when compared to others leading to effective predictions. Figure 6.8 shows the review of ML/DL models employed for effective malware detection. Mostly, Random Forest (RF), Support Vector Machine (SVM) and Naive Bayes (NB) models are adopted for detecting android malware and this also leads to reduced time and space complexity. Other models like Convolutional Neural Network (CNN) and Long Short Term Memory (LSTM) were given next-level priority since they require more computing power.

6.6   CONCLUSION

This work presents a systematic review of malware detection and analysis approaches using ML/DL models. About 42 research papers were reviewed for handling malware analysis and detection using suitable ML/DL models. Those works are reviewed based on various classification criteria like feature sets, ML/DL models adopted, detection approaches, and their solutions and drawbacks. Apart from all these, this work provided a comprehensive survey on ML/DL models focused on feature selection, extraction, and dimensionality reduction techniques for the appropriate classification. The malware analysis procedures are categorized into three types, namely static, dynamic, and hybrid analysis techniques. The classification mechanisms following static, dynamic, and hybrid-based are clearly described in the introduction for better understanding.

An extensive literature survey on malware detection via ML/DL models is performed and compared based on two parameters—system architecture and system input. By considering the above parameters, ML/DL models are clustered as follows: (i) ML/DL models that focus on feature extractions influencing the outputs; (ii) approaches scaling the features based on the input dataset; (iii) techniques adopting the API function sequences; and (iv) techniques analyzing instruction sequences in detail via opcode extraction, function headers, etc.

TABLE 6.4

Survey on ML/DL Models used for Malware Detection and Analysis

References	Detection Approach	Dataset Used	ML/DL Algorithms	Model Accuracy	Strengths	Drawbacks
Lee et al. (2011)	Adopted matrix model implementing Waxshell Algorithm and codes are extracted and represented via API call graph technique	MalGenome	Waxshell extended algorithm	87.75%	False alarms are generated for misclassifications.	Model needs to be expanded for accurate results.
Wang et al. (2016)	Logic design and modeling based on symbols and semantics were derived for android attack	Drebin	C4.5, NB, Linear SVM, RF	97%	Digital Signature Algorithm (DSA) and ML integration increases accuracy.	Static analysis is not adopted results in failure to detect impulsive malwares.
Meng et al. (2016)	Application behavior is captured by integrating system function and eigenvectors via the Androiddetect mechanism.	Google Play	Application functions decision algorithm	90%	Impulsive malwares are tackled by identifying their origin and also results in high performance.	Does not coverall all type of features in code.
Wei et al. (2017)	Amalgamating static, dynamic, and hybrid malware analysis procedures are used to train hidden Markov models	Harebot, Security Shield, Smart HMM Winwebsec, Zbot, ZeroAccess	Hidden Markov Model (HMM)	90.51%	Uses comparison metrics to evaluate ML techniques	Only selected ML algorithms can be evaluated

(Continued)

TABLE 6.4 (Continued)
Survey on ML/DL Models used for Malware Detection and Analysis

References	Detection Approach	Dataset Used	ML/DL Algorithms	Model Accuracy	Strengths	Drawbacks
Damodaran et al. (2017)	Customized ML methodology called Waffle director is derived	Tencent, YingYongBao, Contagio	Decision Trees (DT), Neural Network, SVM, NB, Extreme Learning Machines (ELM)	97.06%	Strong learning procedures lead to less human interference	Imperfect balance between API calls and Permissions
Sun et al. (2017)	Code Heterogeneity Analysis Framework is proposed to classify Android malwares by smali code intermediate representation	Genome, VirusShare, Benign App	RF, KNN, DT, SVM	False Negative Ration (FNR) – 0.35%, False Positive Rate (FPR) – 2.96%	Detailed fine-grained behavioral analysis is performed on program classification	Cannot be employed on DEX encryption techniques and results in malware detection issues
Tian et al. (2017)	Android data automatic analysis procedure prescribes using a set of Python and Bash scripts.	Andrototal	NB, DT	80%	Model execution is efficient.	Less samples are trained in this model, and doesn't considers call frequency.
Leeds et al. (2017)	N-Gram technique was adopted in TinyDroid Framework for scanning opcodes.	Drebin	RF and Average Precision (AP) in TinyDroid	87.6%	Outputs effective classification and detection in identifying malwares.	Metamorphic malware samples are not included for model training and testing
Chen et al. (2018)	In-built package functions are applied in Smali files via API calls	Drebin, Contagio, Google Play	DT, RF, K-Nearest Neighbor (KNN) algorithm, NB	86.89%	This model performs well on less-size datasets	Limited feature set is considered, so not applicable on dataset having larger feature set

(Continued)

TABLE 6.4 (Continued)
Survey on ML/DL Models used for Malware Detection and Analysis

References	Detection Approach	Dataset Used	ML/DL Algorithms	Model Accuracy	Strengths	Drawbacks
Zhang et al. (2018)	RanDroid Framework was proposed to transform binary vectors, feature extraction using ML techniques.	Drebin	SVM, DT, RF, NBs	97.7%	Dynamic analysis was used to analyze API calls, opcode, naive calls lead to improved accuracy.	More number of feature set is not considered such as control flow graph analysis, naive code analysis, etc.
Koli (2018)	DroidEnsemble model was used to create binary vector and evaluate it using ML techniques.	Google Play, AnZhi, LenovoMM, Wandoujia	SVM, KNN, RF	98.4%	String ensembles are characterized by their static behavior and structural features	Detection does not include encryption, anti-disassembly, kernel level features in the given input
Wang et al. (2018)	ML algorithms were integrated with permission vectors and binary feature extraction procedure for malware analysis.	Drebin	RF, J48, NB, Simple Logistic, BayesNet Tree Augmented Naive Bayes (TAN), BayesNet K2, PolyKernel, SMO NPolyKernel	Static-96%, dynamic-88%	Various ML techniques are employed and compared.	MonkeyRunner tool is used to calculate accuracy of the ML Technique.
Kapratwar et al. (2017)	MaMaDroid model is proposed with the implementation of Hidden Markov model determined by API call sequences.	Drebin, Oldbenign	RF, KNN, SVM	94%	Early learning procedure increases prediction accuracy.	Storage complexity increases while performing classification.

(Continued)

TABLE 6.4 (Continued)
Survey on ML/DL Models used for Malware Detection and Analysis

References	Detection Approach	Dataset Used	ML/DL Algorithms	Model Accuracy	Strengths	Drawbacks
Onwuzurike et al. (2018)	Association rule mining is used to calculate behavior semantic of the malwares.	Drebin, AMD	SVM, KNN, RF	96%	System stability is provided with appropriate feature extraction procedures.	Naive codes classification, encrypted codes, etc are not addressed in this methodology.
Zhang et al. (2019)	Applying ML techniques to extract data features from dex manifest classes.	Drebin, playstore, Genome	KNN, SVM, BayesNet, NB, Logistic Regression, J48, RF	98.7%	Fully automated process for effective malware analysis and detection.	All feature set was not considered in DEX analysis.
Kabakus (2019)	Malware analysis is performed by reverse engineering an Input JSON file and running APK in sandbox environment and extracting the features.	MalGenome, Kaggle, Androguard	SVM, LR, KNN, RF	Static-81.03%, dynamic-93%	Improved dynamic analysis procedure was introduced.	Decreased accuracy in terms of hybrid analysis approach.
Jannat et al. (2019)	TFDroid framework was introduced using FlowDroid Static data flow analysis.	Drebin, Google Play	SVM	93.7%	Data flow graph method was used for function descriptions.	Clustering techniques was not considered in designing ML Models.
Lou et al. (2019)	ML/DL models were designed to extract 261 features in the dataset using a hybrid analysis procedure.	MalGenome, Drebin, CICMal-Droid	SVM, KNN, RF, DT, NB, multilayer perceptron (MLP), Guassian Bayes (GB)	99.36%	Improved accuracy was achieved via hybrid analysis compared to static analysis and dynamic analysis.	Not suitable for Runtime Environments.

(Continued)

TABLE 6.4 (*Continued*)
Survey on ML/DL Models used for Malware Detection and Analysis

References	Detection Approach	Dataset Used	ML/DL Algorithms	Model Accuracy	Strengths	Drawbacks
Hadiprakoso et al. (2020)	Static and dynamic features are related using their conditional dependencies. Then, TAN was derived by training Ridge regularized Logistic Regression Classifiers.	Drebin, AMD, AZ, Github, GP	TAN	97%	Improved prediction accuracy.	Impulsive malwares left unidentified.
Surendran et al. (2020)	DLDroid framework was employed to extract features and rank them from Log files using DunaLog tool.	Intel Security	NB, Simple Linear Regression SVM, J48, RF, DL	99.6%	Improved accuracy; framework was tested in real-time environments.	Intrusion detection was not included as a part of malware detection.
Alzaylaee et al. (2020)	Malicious apps were identified using hybrid analysis by incorporating information fusion and Case-based Reasoning (CBR).	Drebin	CBR, SVM, DT	95%	Less memory and processing complexity.	Knowledge representation is missing to produce malware details.
Qaisar and Li (2021)	CNN was employed to analyze features on program codes using opcode filtering.	Drebin, AMD	CNN	91% and 81% on two datasets	Impulsive malwares can be identified by training the model with increased samples.	CNN is not compared with existing ML and DL models.

(Continued)

TABLE 6.4 (Continued)
Survey on ML/DL Models used for Malware Detection and Analysis

References	Detection Approach	Dataset Used	ML/DL Algorithms	Model Accuracy	Strengths	Drawbacks
Millar et al. (2021)	Malwares are detected via suitable ML/DL models by selecting trained feature sets.	Android permissions dataset, computer and security dataset.	Farthest first clustering, MLP, non-linear ensemble decision tree forest, DL.	98.8%	Novel malwares can be identified with higher accuracy.	Dataset types has to be mentioned and needs human intervention in unknown dataset types.
Mahindru and Sangal (2021)	AdMat framework was proposed for malware detection, works based on inputs considered as images and represented using adjacency matrix.	Drebin, AMD	CNN	98.2%	Improved prediction accuracy and its effectiveness.	Limited number of features are used in the defined framework.

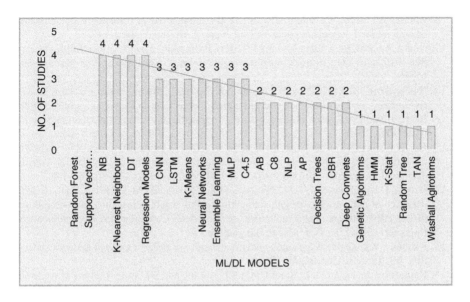

FIGURE 6.8 Visualizing the Studies made on ML/DL models

Additionally, this work introduces research directions on existing classifiers leading to effective malware detection and analysis. Most researchers are still facing many issues related to newly raised malwares, and their effects were also discussed. More emphasis is given to adversarial learning techniques with their drawbacks. Moreover, validation metrics are also presented to evaluate the models' performance based on benchmarks defined by the scientific community.

REFERENCES

1. Alzaylaee, M.K.; Yerima, S.Y.; Sezer, S. (2020), DL-Droid: Deep learning based android malware detection using real devices. Comput. Secur., 89, 101663.
2. Chen, T.; Mao, Q.; Yang, Y.; Lv, M.; Zhu, J. (2018), TinyDroid: A lightweight and efficient model for Android malware detection and classification. Mob. Inf. Syst., 2018, 4157156.
3. Chernis, B.; Verma, R. (2018), Machine learning methods for software vulnerability detection. In Proceedings of the Fourth ACM International Workshop on Security and Privacy Analytics, Tempe, AZ, USA, 21 March 2018. ACM, New York, pp. 31–39.
4. Damodaran, A.; Di Troia, F.; Visaggio, C.A.; Austin, T.H.; Stamp, M. (2017). A comparison of static, dynamic, and hybrid analysis for malware detection. J. Comput. Virol. Hacking Tech., 13, 1–12.
5. Dinaburg, A.; Royal, P.; Sharif, M.; Lee, W., (2008). Malware analysis via hardware virtualization extensions. In: Proceedings of the 15th ACM Conference on Computer and Communications Security. ACM, New York, pp. 51–62.
6. Faruki, P.; Laxmi, V.; Gaur, M.S.; Vinod, P., (2012). Mining control flow graph as api call-grams to detect portable executable malware. In: Proceedings of the Fifth International Conference on Security of Information and Networks. ACM, New York, pp. 130–137.
7. Galal, H.S.; Mahdy, Y.B.; Atiea, M.A., (2016), Behavior-based features model for malware detection. J. Comput. Virol. Hacking Tech., 12 (2), 59–67.

8. Ghiasi, M.; Sami, A.; Salehi, Z., (2015). Dynamic VSA: A framework for malware detection based on register contents. Eng. Appl. Artif. Intell., 44, 111–122.

9. Gupta, A.; Suri, B.; Kumar, V.; Jain, P. (2021), Extracting rules for vulnerabilities detection with static metrics using machine learning. Int. J. Syst. Assur. Eng. Manag., 12, 65–76.

10. Hadiprakoso, R.B.; Kabetta, H.; Buana, I.K.S. (2020), Hybrid-based malware analysis for effective and efficiency android malware detection. In Proceedings of the 2020 International Conference on Informatics, Multimedia, Cyber and Information System (ICIMCIS). IEEE.

11. Hu, X.; Shin, K.G.; Bhatkar, S.; Griffin, K., (2013). MutantX-S: scalable malware clustering based on static features. In: Presented as Part of the 2013 USENIX Annual Technical Conference (USENIX ATC 13). USENIX, San Jose, CA, pp. 187–198.

12. Jannat, U.S.; Hasnayeen, S.M.; Shuhan, M.K.B.; Ferdous, M.S. (2019) Analysis and detection of malware in Android applications using machine learning. In Proceedings of the 2019 International Conference on Electrical, Computer and Communication Engineering (ECCE). Cox'sBazar, Bangladesh.

13. Kabakus, A.T. (2019) What static analysis can utmost offer for android malware detection. Inf. Technol. Control, 48, 235–249.

14. Kapratwar, A.; Di Troia, F.; Stamp, M. (2017), Static and Dynamic Analysis of Android Malware. ICISSP, Porto, Portugal, pp. 653–662.

15. Kim, S.; Yeom, S.; Oh, H.; Shin, D.; Shin, D. (2021) Automatic malicious code classification system through static analysis using machine learning. Symmetry, 13, 35.

16. Koli, J. (2018), RanDroid: Android malware detection using random machine learning classifiers. In Proceedings of the 2018 Technologies for Smart-City Energy Security and Power (ICSESP). IEEE, pp. 1–6.

17. Lee, J.; Im, C.; Jeong, H., (2011). A study of malware detection and classification by comparing extracted strings. In: Proceedings of the 5th International Conference on Ubiquitous Information Management and Communication. ACM, New York, p. 75.

18. Leeds, M.; Keffeler, M.; Atkison, T. (2017), A comparison of features for android malware detection. In Proceedings of the SouthEast Conference, Kennesaw, GA, USA, ACM, New York, pp. 63–68.

19. Lou, S.; Cheng, S.; Huang, J.; Jiang, F. (2019) TFDroid: Android malware detection by topics and sensitive data flows using machine learning techniques. In Proceedings of the 2019 IEEE 2nd International Conference on Information and Computer Technologies (ICICT), IEEE. pp. 30–36.

20. Luo, L.; Dolby, J.; Bodden, E. (2019) MagpieBridge: A general approach to integrating static analyses into IDEs and editors (tool insights paper). In Proceedings of the 33rd European Conference on Object-Oriented Programming (ECOOP 2019). Schloss Dagstuhl-Leibniz-Zentrum fuer Informatik, Dagstuhl, Germany.

21. Mahindru, A.; Sangal, A. (2021) MLDroid—framework for Android malware detection using machine learning techniques. Neural Comput. Appl., 33, 5183–5240.

22. Meng, G.; Xue, Y.; Xu, Z.; Liu, Y.; Zhang, J.; Narayanan, A. (2016), Semantic modelling of android malware for effective malware comprehension, detection, and classification. In: Proceedings of the 25th International Symposium on Software Testing and Analysis. ACM, New York, pp. 306–317.

23. Millar, S.; McLaughlin, N.; del Rincon, J.M.; Miller, P. (2021), Multi-view deep learning for zero-day Android malware detection. J. Inf. Secur. Appl., 58, 102718.

24. Mohaisen, A.; Alrawi, O.; Mohaisen, M., (2015). Amal: High-fidelity, behavior-based automated malware analysis and classification. Comput. Secur., 52, 251–266.

25. Nataraj, L.; Karthikeyan, S.; Jacob G.; Manjunath, B.S., (2011). Malware images: Visualization and automatic classification. In: Proceedings of the 8th International Symposium on Visualization for Cyber Security. ACM, New York, pp. 1–7.

26. Onwuzurike, L.; Mariconti, E.; Andriotis, P.; Cristofaro, E.D.; Ross, G.; Stringhini, G. (2018) MaMaDroid: Detecting android malware by building Markov chains of behavioral models (extended version). ACM Trans. Priv. Secur., 22, 14.

27. Pang, Y.; Xue, X.; Wang, H. (2017), Predicting vulnerable software components through deep neural network. In Proceedings of the 2017 International Conference on Deep Learning Technologies, Chengdu, China, 2–4 June 2017. ACM, New York, pp. 6–10.

28. Ponta, S.E.; Plate, H.; Sabetta, A.; Bezzi, M.; Dangremont, C. (2019) A manually-curated dataset of fixes to vulnerabilities of open-source software. In Proceedings of the 2019 IEEE/ACM 16th International Conference on Mining Software Repositories (MSR), Montreal, QC, Canada, 26–27 May 2019. IEEE, pp. 383–387

29. Abusnaina et al., (2021), Systemically Evaluating the Robustness of ML-based IoT Malware Detectors,*2021 51st Annual IEEE/IFIP International Conference on Dependable Systems and Networks - Supplemental Volume (DSN-S)*, 3–4, doi: 10.1109/DSN-S52858.2021.00012.

30. Qaisar, Z.H.; Li, R. (2021), Multimodal information fusion for android malware detection using lazy learning. Multimed. Tools Appl., 81, 12077–12091.

31. Russell, R.; Kim, L.; Hamilton, L.; Lazovich, T.; Harer, J.; Ozdemir, O.; Ellingwood, P.; McConley, M. (2018), Automated vulnerability detection in source code using deep representation learning. In Proceedings of the 2018 17th IEEE International Conference on Machine Learning and Applications (ICMLA), Orlando, FL, USA, 17–20 December 2018. IEEE, pp. 757–762.

32. Sun, Y.; Xie, Y.; Qiu, Z.; Pan, Y.; Weng, J.; Guo, S. (2017), Detecting Android malware based on extreme learning machine. In Proceedings of the 2017 IEEE 15th International Conference on Dependable, Autonomic and Secure Computing, 15th International Conference on Pervasive Intelligence and Computing, 3rd International Conference on Big Data Intelligence and Computing and Cyber Science and Technology Congress (D ASC/PiCom/DataCom/CyberSciTech). IEEE, pp. 47–53.

33. Surendran, R.; Thomas, T.; Emmanuel, S. A (2020), TAN based hybrid model for android malware detection. J. Inf. Secur. Appl., 54, 102483.

34. Tahaei, M.; Vaniea, K.; Beznosov, K.; Wolters, M.K. (2021), Security notifications in static analysis tools: Developers' attitudes, comprehension, and ability to act on them. In Proceedings of the 2021 CHI Conference on Human Factors in Computing Systems, Yokohama, Japan, 8–13 May 2021. ACM, New York, pp. 1–17.

35. Tian, K.; Yao, D.; Ryder, B.G.; Tan, G.; Peng, G. (2017), Detection of repackaged android malware with code-heterogeneity features. IEEE Trans. Dependable Secur. Comput., 17, 64–77.

36. Vu, L.N.; Jung, S.(2021) AdMat: A CNN-on-matrix approach to android malware detection and classification. IEEE Access, 9, 39680–39694.

37. Wang, W.; Gao, Z.; Zhao, M.; Li, Y.; Liu, J.; Zhang, X. (2018), DroidEnsemble: Detecting Android malicious applications with ensemble of string and structural static features. IEEE Access, 6, 31798–31807.

38. Wang, Z.; Li, C.; Yuan, Z.; Guan, Y.; Xue, Y., (2016), DroidChain: A novel Android malware detection method based on behavior chains. Pervasive Mob. Comput., 32, 3–14.

39. Wei, L.; Luo, W.; Weng, J.; Zhong, Y.; Zhang, X.; Yan, Z. (2017). Machine learning-based malicious application detection of android. IEEE Access, 5, 25591–25601.

40. Zhang, H.; Luo, S.; Zhang, Y.; Pan, L. (2019) An efficient android malware detection system based on method-level behavioral semantic analysis. IEEE Access, 7, 69246–69256.

41. Zhang, P.; Cheng, S.; Lou, S.; Jiang, F. (2018), A novel Android malware detection approach using operand sequences. In Proceedings of the 2018 Third International Conference on Security of Smart Cities, Industrial Control System and Communications (SSIC). IEEE, pp. 1–5.

7 Healthcare System 4.0 Driven by Quantum Computing and Its Use Cases
A COVID-19 Perspective

Naga Raju Mysore
Presidency University

CONTENTS

DOI: 10.1201/9781003250357-7

7.1 INTRODUCTION

In healthcare systems nowadays, in addition to physical form, digital versions and registries of diseases, health records of patients, drug research, and drug development and their clinical trials information are included. The Compound Annual Growth Rate (CAGR) [1] of the global digital healthcare facility is estimated to be 22.4% (shown in Figure 7.1).

Such a huge growth can only be better managed by computing systems than manual. The main aspects of computerized hospitals are the evolution of medical informatics and pharmacoinformatic, from the cognitive underpinnings of medicine, pharmacy, and computer science [2]. These systems manage huge data effectively, enforce stringent electronic data exchange standards, handle drug aspects efficiently, and perform the role of expert systems to help medical professionals to diagnose and treat certain critical diseases very effectively.

Other factors of a computerized hospital that leave an impact on patients are a drastic reduction in patient wait time, fewer mistakes in patients' treatment, lowering hospital-acquired infections by patients, providing better healthcare experience and cheaper treatment cost, offering better care, facilitating quick diagnosis of the ailment of the patient and timely treatment of it [3]. Further, computerized hospitals are capable of storing a large amount of patient data with a facility to access and share that data among medical professionals in a secured manner. They can also help medical professionals in predicting disease trends clearly and provide preventive care for the patient.

On 11 March 2020, the rapid increase in the number of cases outside China led the WHO (World Health Organization) Director-General to announce that the outbreak could be characterized as a pandemic [4]. By then, more than 118,000 cases had been reported in 114 countries, and 4,291 deaths [4] had been recorded. This pandemic

Worldwide Digital Healthcare Size in US Dollars (in Millions)

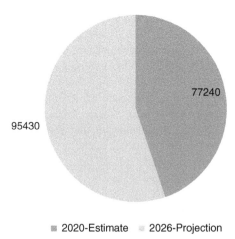

FIGURE 7.1 Worldwide digital healthcare market size in US dollars (million)

has thrown challenges to issues like the health of people all over the globe and the ill impact of the virus on societal and economical aspects, specifically on low- and middle-income populations of developing and poor countries with fragile health systems. The virus menace forced for travel, movement, and social restrictions all over the globe. This situation demanded for identifying and isolating virus-affected areas as clusters and imposing strictly COVID-appropriate behaviour rules there to avoid further spread. This further necessitated public-health measures like close surveillance, quick identification, and community spread avoidance. In 2003, Hong Kong [5] successfully managed the outbreak of earlier class of Severe Acute Respiratory Syndrome by adopting electronic data systems. From 2014 to 2016, West Africa managed the Ebola virus outbreak effectively using mobile phone data [6]. In a similar manner, many digital technologies are developed and put into use to take public-health measures mentioned above under the COVID-19 situation [7].

The behaviour of patients towards seeking in-person medical care has dropped to the lowest level during the pandemic, mainly due to fear of catching COVID-19. Many of them cancel in-person appointments and postpone routine tests due to the same fear [8]. This study shows avoidance of in-person medical care and in turn substantial growth in demand for telehealth on large scale. Figure 7.2 depicts this information collected from the National Health Service in the UK for the year 2019.

To study and analyse drug interaction completely, many research industries are spending huge amounts. In 2018, Pharmaceutical Committee reported that above 26 million people in England suffer from long-term illness. According to data published by National Health Service during 2019, 64%, 50%, and 70% accounted for outpatient, general practitioner, and inpatient bed days [9]. Growing demand for speciality health consulting in remote areas has boosted the adoption of digital health. Health records kept in digitized form facilitate greater assistance to medical practitioners to do disease diagnosis and prescribe treatment to the patient very effectively.

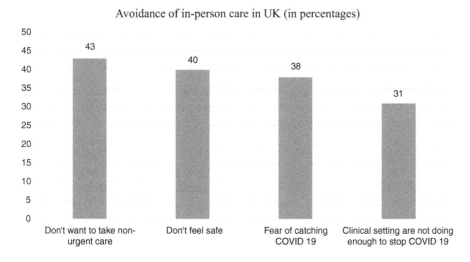

Avoidance of in-person care in UK (in percentages)

FIGURE 7.2   In-person care avoidance of other ailment patients in the UK due to COVID-19 pandemic risk based on data collected from NHS for the year 2019

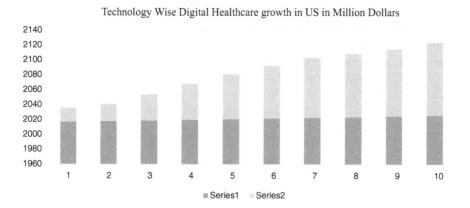

FIGURE 7.3 Growth of digital healthcare technologies in the US in million dollars

A few technologies encompass the digital health market. They are Tele-healthcare, mhealth, Healthcare Analytics, and Digital Health Systems. All these definitely grew during this decade as their relevance and importance got noticed due to the COVID-19 pandemic situation. The predicted growth of these technologies in the US is shown in Figure 7.3 given below. In this figure, series 1 and series 2 represent tele-medicine and mhealth technologies of the healthcare system, respectively.

Artificial Intelligence (AI)/Machine Learning (ML) is one technology touching all facets of the healthcare system. This technology is used to identify new molecules to predict adverse events to suggest the best treatment pathways. The usage of AI/ML in healthcare is very diverse and deep. Genetic and deep learning techniques coupled with advanced options like QC are going to further open new doors [10].

Digital health technology, especially QC adoption, has been crucial during this time — relieving pressure on healthcare systems and enabling the continuation of healthcare for many in a virtual manner in the form of telemonitoring, apps, and AI. The adoption of QC by the healthcare system updates it to version 4.0 which is definitely going to serve fourfold objectives, namely, detecting disease very precisely and quickly thus paving the way for excellent healthcare, allowing treatment of ailments at an early stage and at a cheaper cost, strengthening patient experiences much towards reality, and upgrading medical professional technical practices to new approaches. The upgraded medical professional technical practices include silico trials on simulated humans that replace clinical trials on actual humans. It will cut down the risk involved in human trials to a minimum and facilitate coming up with the required drug with supersonic speed. It leads to hospital on cloud. Under a pandemic situation like COVID-19, such a system has got great relevance, wide application, and usage. COVID-19 drug discovery can be accelerated, automated, and done in an economical manner using quantum machine learning, a new field which is the combination of machine learning and quantum mechanics [11]. Lack of preparedness has been seen in the area of drug discovery. To devise an effective drug quickly enough, we need to do a lot of research and this drug research is a computationally-intensive task. We need to analyse the various kinds of interactions between the biomolecules and which ingredients are to be taken in what proportion. In order to tackle infectious agents such as viruses, there can be no other better way than to model the problem on a computer and conduct extensive research on the same. In the quest of finding a drug for coronavirus that prevents the virus from affecting human cells in a quick manner, the real need was to analyse a large number, nearly 8,000 compositions, in order to know those compounds that stick to its spike the most. This demanded extraordinary computational power. IBM deployed the world's largest supercomputer, Summit, to assist in the quest. It allowed researchers to execute a huge number of computations involved in atomology and bionomics through their simulation models successfully. After that, they recommended 77 drug compositions that could be the candidates for clinical tests. This is how drug research can be accelerated through the high and fast computational power of QC systems.

One realization that the world has had with this pandemic is that the computing power and paradigm need to evolve to solve complex problems which even the best supercomputers find tedious to do. The world can't afford to lose a battle against the global risk of infectious diseases due to the lack of adequate computing power. Quantum computers offer the promise of bridging the large computational capability gap. Drug companies require over 10 years of time and billions of dollars of investment for new drug research until brought to market. They also face a high risk of failure in drug research after their long effort due to the lack of conducting experiments very precisely. Upgrading computing systems by quantum aspect cuts cost by a substantial amount and propel time to market the drug to the quickest possibility. Further such systems enable chemists to discover drug compositions for a wide range of diseases very fast and effectively and allow them to study the characteristics of such compositions very precisely.

QC evolves healthcare system to version 4.0. Such a system will be able to provide authoritative data with robust vision at the beginning stage of the ailment itself to medical consumers. It facilitates medical consumers to get valuable insights from their health data at the early stage of the ailment, and they can be in a position to take better decisions regarding treatment due to the aspect of quick diagnosis through QC. Also, based on this, the medical consumer comes to know about the life-threatening ailment, if any, at the early stage itself before it goes to the worst stage. As a result of that, the patient can be treated at the initial stage of the ailment leading to his/her life saving as well as incurring cheaper treatment costs. This aspect of healthcare system 4.0 can be termed predictive healthcare. A medical system with QC can also do fast genome sequencing. Further, patients' digital records can be ultra-secure through the QC uncertainty aspect [12]. The underlying requirement of such a medical system is upgrading its competence from traditional computing to the next level, like, QC. QC not only speeds up things exponentially but also necessitates diverse intelligence, a fresh and hugely desired new skillset, noticeable IT architectures, and innovative corporate approaches. It shields sensitive medical data of the patients in an ultra-tough secured manner from the access of intruders.

QC will bring in considerable breakthroughs in the healthcare system. These are discussed as follows. It helps in having an optimal radiation plan in radiotherapy, that kills cancer cells very precisely while keeping damage to surrounding tissues the least. It quickens drug research by saving time as well as the amount invested to a considerable level. It facilitates modelling of molecular interaction of proteins and understanding of human genome encoding well and fast in order to produce new drugs. It also makes research of diagnosis and screening of disease capable to study huge complex patient data effectively and identify disease in a correct manner, respectively, by excellent pattern recognition facilities. The artificial intelligence of machines can be enhanced remarkably by QC by launching quantum machine learning as a new field. It also revolutionizes medical imaging through quantum sensors [13], so that Magnetic Resonance Imaging images be done to a more basic and clear level. This provides clinicians with more accurate images in order to study the ailment of the patient very effectively. Moreover, it also facilitates fast DNA sequencing due to its ability to deal with huge data and be able to perform complex computations. This will open up personalized medicine. These computers are also able to unfold protein more accurately which is not yet well understood to date. This will result in new therapies.

While the bright side of powerful QCs will solve a lot of problems in humanity and will give a huge boost to the discovery of drugs, new materials, and space research, the dark side of QCs at the same time are cyber-attacks, data theft or fraud, and breakdown of information infrastructure. Clients in major economies are approaching companies with quantum cryptography capabilities to help them start their journey towards a quantum safe future. The emerging need of the world is a secure networked digital economy and quantum technology plays a pivotal role in fulfilling this need. Quantum mechanics-based computers handle huge patient data definitely well during all its phases like storage, transmission, and security [14].

The main contribution of this work is to present the objectives drawn from the usage of the underlying features such as qubits, superposition, entanglement, and uncertainty of QC in the area of healthcare. First, QC systems provide high and speedy computing resources. Second, they replace clinical trials with silico trials. Third, genome sequencing can be done in a very precise manner and results can be obtained in a very short period of time. Finally, it will provide for things like predictive healthcare analysis by which medicines can be made for patients at an early stage of their disease, process automation, and high security of medical data.

These objectives will result in observations such as speeding up the vaccination development process, avoiding all risks involved in human trials completely, and the efficacy of the vaccine can be tested by conducting any number of trials with no human risk. Besides, medical practitioners can perform patient-tailored diagnosis and treatment. In the end, QC-based healthcare system updates it to version 4.0 and keeps itself ready to manage future situations better.

This chapter is organized as follows. This section dealt with the introduction part. Section 7.2 includes literature review by including illustration of its preliminaries. Section 7.3 lists new and possible use cases in healthcare system 4.0. Section 7.4 outlines the future scope. Finally, Section 7.5 presents the conclusion.

7.2 LITERATURE REVIEW

7.2.1 Preliminaries

New virtualized business models of healthcare with more efficiency and effectiveness could be evolved through QC. Such systems improve and streamline medical operational processes by modernizing them. They also provide for more customer intimacy in an easy way. They further comply with medical regulations very well. QC gives strategic and economic advantages besides ultrafast processing capacities. Indian government recently launched National Mission on Quantum Technologies and Applications (NMQTA) by allocating a budgeted amount of ₹8,000 crore for the financial year 2020–2021 [15]. This is completely meant for the development of QC-linked technologies. With this initiative, India occupies the globally third position in this section after the US and China.

7.2.1.1 Quantum Computing Insights

- QC is drawn from quantum mechanics principles. It has been developed in the twentieth century beginning in order to study and express things at a small scale like atoms and a basic level of particles.
- Traditional computers store data and solve problems using either zero or one bit. Quantum computers use quantum bits called qubits that are typically used to represent electrons and photons which can exist as both zero, one, or zero and one at the same time.
- The concept of superposition gives the capability that the qubit can be in numerous states simultaneously. Another concept called entanglement enables the coexistence of two paired qubits during each state to allow performing many computations parallelly as well as providing for mighty computational ability.
- Simulations that involve complex computations with a huge blend of divergent variables and high processing power requirements could be handled better by quantum computers than traditional ones. This feature of QC would be very much suitable for the healthcare system.
- In a short span of time from now, in the medical field, QC could help in choosing the correct patient and in clinical trials design, in the fast development of state-of-the-art molecules with a coveted set of biological compositions, in the superior prediction of drug response and its fast marketing, and in developing new drugs for diseases that cannot be treated at present.
- QC has the ability to bring in silico trials in place of clinical trials. This minimizes the risk aspect, facilitates conducting trials as many times as required to achieve efficacy of the medicine, and saves a lot of time and money in research by pharmaceutical companies. This facilitates drug makers to come up with new drug quickly and bring it to commercial use in a very short amount of time.
- QC-administered healthcare helps both medical companies and their clients. It brings in the quick administering of patient's admission, beds allotment, preparation of treatment schedule, avoiding needless diagnostic testing, and MRI imaging optimization. Deep learning clubbed with QC

leads to Generative Adversarial Networks (GANs). They could be capable of handling sparse image data for uncommon diseases.

7.2.1.2 Challenges Associated with Quantum Computing

- The first threat is to have a quantum superposition of qubits in a greatly restrained way. Qubits have lax "quantumness" if not restrained correctly due to their brittle nature. Apart from this, to get them work conscious materials preference, plan, and building are mandatory.
- The second one is the challenge of creating algorithms and applications for quantum computers.
- The third one is the demand for the development and control of hardware as well as software of QC in an altogether new way.
- Fourth, the security of quantum-based systems should be handled differently due to their complex and fast computational nature.

7.2.1.3 India's Effort and Initiative

- QC research got initiated in the global arena two decades back. In India, this has been started around 5 years back.
- Indian government started taking serious note on QC only during the year 2018. It initiated around 51 institutions in this area under Quantum Enabled Science and Technology. No major advancement occurred until the constitution of NMQTA [16].

7.2.1.4 Advancement of Quantum Computing in India

The Indian government through the announcement of NMQTA wants to allocate funds in an enormous manner that is matching to funds allocated for the same by the US and Europe. Nonetheless, some important challenges of QC need to be addressed. This needs the following to be pursued seriously:

- Managing a remarkable combination of collaboration among theoretical as well as experimental physicists, computer experts, material analysts, and engineers.
- Indian government wants to have reputed institutions and the scientific community as a part of this mission to come up with a detailed plan and fast implementation of the same.
- Indian government looks for private industry and philanthropists to be a part of this mission for the following two reasons:
  - To make the private sector notice and understand the importance of QC
  - Creating a dynamic intellectual environment to notch up top researchers across all sectors

7.2.2 HEALTHCARE SYSTEM 4.0

QC enables the current medical system to evolve into version 4.0 through the paramount innovations possible. It can revolutionize disease diagnosis in such a healthcare system to a much more precise level. This could be achieved through the invention of quantum imaging devices that use quantum sensors. In imaging, MRI machines equipped with quantum sensors are able to notice things in the image at the molecular level, thus providing more accurate figures and pictures to clinicians facilitating them to analyse them very precisely. In radiotherapy, it could facilitate obtaining the best radiation procedure in order to lessen the harm caused to neighbouring tissues and body units to the lowest level. In drug research, it could shorten drug evolution span considerably and save billions of amounts poured into compound drug design, also propelling the process of examining for cures for a wide range of illnesses. In the diagnostics research and disorder screening process, a medical system equipped with QC will be able to process a huge amount of medical data of the patient to uncover potential disorder patterns. QC can also improve the artificial intelligence of computing machines enabling the automation of many such processes. They lead to new systems termed Quantum Machine Learning Systems. Such systems replace clinical trials with silico trails in order to bring down the risk involved in trials to the lowest level and speed up the trial process. Besides, it results in personalized health care. In genomic medicine, it quickens DNA sequencing procedures and results in the opportunity for personalized medicine. In much-complicated protein folding, it may promise new horizons for novel therapies. Quantum computer potentially assists in better management of healthcare aspects at different stages of data like repository, transmission, and security. QC option is the only one that supports achieving a healthcare system with personally tailored healthcare, genomics dependency, personal genetics, and customized drugs for patients. It helps to diagnose the disease at an early stage. This leads to predictive healthcare. Another aspect is cloud QC. These use cases are briefly presented as follows.

7.2.3 APPLICATION OF QUANTUM TECHNOLOGIES AND COMPUTATION IN HEALTHCARE SYSTEM

In this article, the following use cases of the new healthcare system are considered by assuming them as prominent in nature. The illustrations of these use cases are given in their respective subsubsections as well.

7.2.3.1 Quantum Sensors and Quantum Imaging in Disease Diagnosis

Mainly, sensors' dimensions and imaging quality play a vital role in disease analysis and diagnosis. The following subsections will give a detailed account of these aspects.

7.2.3.1.1 Quantum Sensors

Quantum sensors will bring in a manifold improvement in medical equipment and enable disease diagnosis to happen in a very precise manner. Quantum sensors fitted medical devices will be light, compact, very precise, and easily manageable. They are of sugar cube size. They are so precise that they can make measurements very close to the reality of the quantum world. QC is able to consider the linkage of particles residing at great distances through the concept of entanglement, or particles as well as their reactions could be viewed in two places at once by another concept known as superposition. QC further adds for extensive and complex computations of data sensed. The macQsimal by European Union is one such project. In this project, small quantum sensors are used to enhance the medical imaging process. Enhancement of imaging will lead to better Magnetoencephalography (MEG) scanners [17]. These scanners are bulky and to be cooled by liquid nitrogen or helium. Due to their bulky nature, they cannot reach near that part of the patient to be scanned. As a result, measurements are not so accurate in nature. But small quantum sensors resolve this problem as they can be moved very near to the part of the human to be scanned in order to capture observations precisely. Also, several such sensors can be fitted to the scanner machine and capture measurements at different positions of the scanned part of the human. This brings in very precise brain and cardiac imaging. These precise measurements and observations could lead to better analysis of scanned images. This in turn brings in a better understanding of ailments. This will contribute to much better drug research and development that can come up with the right drug finally [18].

Another useful concept of quantum mechanics is hyperpolarisation, by which scanners can be made more perceptive and correct in nature. Using this concept, EU initiated during October 2018 another project by the name MetaboliQS [19]. The biomolecules are injected into a human so that they can accumulate in certain tissue to be observed. Later, this accumulation helps in MRI scanning process to closely observe what is happening in that tissue in a precise manner. These biomarker molecules provide for very minute observation of activities happening in that tissue. They need to be cooled at $-270°C$ first and then warmed to body temperature at the time of their use. It needs a long time and incurs high costs. Diamonds constructed using quantum sensors require mild cooling or no cooling at room temperature at all. Such sensors allow MRI scanners to observe time-sensitive effects in cancerous tissue more properly. It also allows for having very detailed images of the affected cells. These improved images facilitate identifying earlier, later stages of tissue or dead tissue. These bio-sensors implants enable medical practicians to understand the development of illness in the human body correctly.

The success of these projects will lead to precise observations and effective tackling of many health issues that are not possible to be addressed by traditional computers at present. The diagnosis of complex diseases like Dementia and Alzheimer's become very precise due to MRI scanners with embedded sensors of small diamond shape. This will scale down MRI scanners to nano levels and facilitate the investigation of biosystems that are very small in size. Such technologies bring in a quantum era that brings in remarkable changes in the overall healthcare system.

7.2.3.1.2 Quantum Imaging

Machines with quantum imaging facility would be light and small in size. They make the imaging process very precise in nature. They could also help in interpreting the results of treatment. Automation of machine through machine learning helps in detecting abnormalities of illness of humans, and QC facilitates the right interpretation of results of the treatment of that illness. They led a new research domain by the name of quantum machine learning [20]. Present MRIs identify affected areas not so clearly. Further evaluation is to be done by a radiologist. But it reduces that burden on the radiologist to a greater extent making the machine do that job on its own to the maximum extent possible.

7.2.3.1.3   Quantum Radiotherapy

It is a cancer treatment. It streams radiation beams on cancer cells. They are used to either kill cancer cells or stop their multiplications. They make such cells die or slow down their multiplications by damaging the DNA of tumour cells [21]. Quantum radiotherapy facilitates to have an optimal radiation plan that can stream radiation beam only on cancer cells by not affecting goods cells surrounding it. High doses of radiation are very harmful and lead to severe health-damaging consequences later in the patient if not used properly. Achieving such an optimal radiation plan is more complicated in nature as it involves thousands of variables associated and their extensive computations [22]. An optimal radiation plan is achieved only after a large number of simulations on traditional computers. Such a plan requires a huge amount of time. On the other hand, quantum computers facilitate parallel running of large numbers of simulations and thus allowing to determine optimal plan in a short amount of time [23]. This feature attracts many medical research companies. The lack of these features in classical computer systems makes them unsuitable for such a task. QC based systems will be an ideal choice for such a requirement as it provides for a much broader horizon for considering any number of possibilities between each simulation, facilitating simultaneous execution of multiple simulations and finally leading to come up with an optimal plan very quickly. Its market was 93 million dollars in 2019 and will reach 283 million dollars by 2024 [24]. The first factor of QC that propels its growth is the supersonic speed discovery of drugs for new ailments. The second factor for its growth is the substantial reduction in investments for drug research. The third factor contributing to its growth is facilitating a comprehensive analysis of the disease.

The biggest challenges in administering ionizing radiations in cancer patients are to give optimal dose targeting only tumours and keeping that radiation effect as minimum as possible on normal cells. Intensity-Modulated Radiation Therapy (IMRT) is one of the latest methods of radiation therapy [21]. In this therapy, the radiation applied is adjusted to get to an ideal dose to administer inside the patient. The achievement of the ideal dose is dependent on exploring the extremely important right way handling of the mathematical problem with a huge amount of optimization parameters and a large number of constraints. They are supported by remarkable features of quantum computation. In spite of this, its application is still limited. Lack of hardware for quantum systems as well as unavailability of direct mapping of classical to quantum scenario hindering its usage in various applications. To beat these drawbacks and to adopt quantum motivated methods in solving optimization problems, a further level of development is to be carried out in quantum technologies.

For example, consider applying Tensor Networks (TNs) to IMRT [22] ideal radiation measure achievement. Cost function is devised and then optimized by a lengthy range of spirals using TNs. IMRT is one of the radiotherapy methods. Tumour is treated by projecting a photon beam from various angles $(\theta_1, \ldots, \theta_N)$. Grid is formed in order to subdivide photon beam into beamlets. These beamlets are modulated through weights accordingly. Quantification of each beamlet contribution to the whole beam is denoted by weight $x_j \geq 0$ for j-th beamlet. These weights serve as variables to be used to optimize cost function $F(x_1, \ldots, x_{NB})$. Each weight represents the difference between the required and given dose. The ultimate objective is to come up with optimized beamlets with weights that achieve cost function minimization.

This minimization is achieved through the IMRT method by employing TNs. Weights, x_j, represent the global range of interacting spins of pairs. Minimizing cost function is made through Ising-like Hamiltonian. Involving tree tensor network algorithm. The ultimate composition of spins leading to optimal radiation plan is arrived at by re-adjusting beamlets weights accordingly.

It is also noted that the results of TNs are quite compatible for conducting Quadratic Programming (QP) and Simulated Annealing (SA) on them. It is also further observed that these methods are applicable to highly complicated and pragmatic medical cases of similar types of medical computations.

7.2.3.2 Quantum Computing Impact on Drug, Diagnostic Research, and Disorder Screening Process

A quantum computer could be used to optimize drug design and the drug testing process. Quantum computers can also perform simulations and could compute accurate simulations of a new drug on virtual human subjects, only within a few hours. This would save drug companies money and time, as well as remove the number of test subjects for a study, be it animal or human test subjects. Using quantum computers can speed up the drug design and test process, offering new medicines that could save potentially thousands of lives. For drug companies, this would save them thousands, if not millions, in years of drug testing and drug development. It could be used to improve patient choice for trials, conduction of analytical trials, fast production of the new component with required biological aspects, elegant prediction of drug response and its quick marketing, and cater to treat different diseases that are not treated yet.

Drug research at present depends on computerized models and simulations (M&S) to study the nature of atoms and molecules and to come up with new drugs that show more good effects and least damaging ones. However, these models and simulations are facing utilization limits in the form of being unable to observe complicated and highly processing oriented activities of a molecule. The objective of such a process is to develop the composition of molecules showing the maximum endurance of composition at its ground state. Modelling and simulation tools in Ref. [23] help drug analysts to achieve this objective by facilitating to model interactions among each compound electrons, to closely study the interaction among them. The task will be direct if the molecules tested involve simple interactions among them. However, that is not the case as they involve very complex interactions. Present high-end computers can simulate the chemical activities of few numbers of atoms and molecules. Computations will be very narrow and short in portions. New and life-saving drugs need a large number of molecules with complex activities to be observed with extensive computations that traditional computing systems are not capable of. As a result, in clinical trials, researchers are compelled to consider an approximate study of molecular reactions to understand their reactions. It is highly incompetent as the majority of drugs do not succeed in the first phase of clinical trials. This causes high failure rates in drug development. In addition, traditional computing is unable to capture quantum interactions that clinch the actual essence of a molecule. These factors obviously expose the long-overdue in drug discovery technological updates.

Lack of achievement of efficacy is the main issue in traditional computers. Efficacy can be specified as determining the most favourable result among tested ones which require high-intensity computations. The right choice for such a requirement is quantum computers. They provide for different states of existence considered parallelly and are able to capture interactions among chemical compounds at basic particle levels. They are fast and also provide a high capacity of processing power. They bring in some problems in the form of errors. Kandala at MIT Review's 35 Innovators Summit showed how its errors can be managed to improve quantum computations accuracy [24]. This attempt promises practical use of it by industry in the short amount of time possible.

Its fast and intensive processing abilities can be used in making an exact diagnosis. It also helps in propelling machine intelligence to the extent that can serve as aid and also lessen the burden on the study of the process by humans a lot. It converts an MRI scanner into a high-quality one that produces very precise images enabling a detailed level of information to clinicians while screening the ailment. Through this, a doctor can understand a patient's disease very well. Based on this information, personalized treatment can be provided to a patient with respect to age, race, gender, genetics, etc. Virtual simulators [25] are being developed to carry out silico trails instead of clinical trials. They shorten the time and avoid risk. Hundreds of combinations of compounds can be tested and come up finally with the right drug.

7.2.3.3 Quantum Machine Learning Systems

These systems are produced by the combination of concepts like machine learning and QC. It enhances the artificial intelligence of machines to the next level. It provides precise and fast computational capabilities. These features enable the machine to get trained first in analysing the task and later do such task on its own relieving that burden from humans. However, it requires the development of new algorithmic approaches for the task. It needs the development of new hardware like Application Specific Integrated Circuits (ASICs) and Field-programmable Gate Arrays (FPGAs) [26] that are more specific in functionality. They also require Graphics Processing Unit (GPU) with great computational capabilities. These systems' underlying architecture is made up of neural networks and similar ones. These systems would be able to handle electromagnetic and laser pulse strengths for observing as well as solving the problem.

Features like Differentiable Programming and Deep Learning are required for realising quantum machine learning aspects. PyTorch and TensoFlow software libraries are needed to carry out such features. Differentiable Programming is enhanced than Deep Learning. Differentiable Programming feature is one in which algorithms are not coded but rather learned by the machine. This is how trainable quantum circuits of the quantum computer get trained. This will help in developing quantum algorithms, correcting quantum errors, and understanding such systems [27].

7.2.3.4 Silico Trials and Personalized Health Care

The problems in clinical trials are long duration and huge investment with no guarantee of the right drug evolution after such efforts. They involve the risk of carrying out tests on animals and humans. In silico trials, virtual simulators replace animals and humans reducing risk to a minimum. These systems enable researchers to carry out various possible options and obtain results very fast. In silico trials, human organs on chips are used for research. Researchers of the Wyss Institute have been working on the first strand, human organs on chips for years. These microdevices are developed online with human organs. They can mimic the functionalities of the corresponding living organ very well. In silico trials, living organs are not required; rather, virtual organs of humans are enough to consider to perform tests. In these trials, researchers can be provided with a simulated model with a visualization facility of biological processes. They can conduct tests in a virtual manner, observe them, and come up with treatment. Simulations are the "heart" of the silico trials. In these systems, researchers can try all options in a quick manner to know the right drug to treat it well. Such systems also provide for personalized treatment based on age, gender, and many other aspects of the patient as drug reactions will be different for different patients. More effective diagnosis and treatment of the individual patient is made possible. Cognitive computing with bioinformatics makes simulated models to process biological things well that too under virtual setup with zero risk of life endangerment. The production of results in a quick amount of time and offering more options to try are favourable factors driving towards the high adoption rate of such systems. AI boosted by it makes such systems able to find out patterns and clusters clearly out of any unstructured data with great precision in a quick amount of time [28]. They avoid adverse events like patient dropout and compliance with the treatment. Thus, hurdles in trails can be removed to the maximum extent. One such simulated system is HumMod which considers large numbers of aspects like body fluids, circulation, electrolytes, hormones, metabolism, and skin temperature of each patient. This brings in the realization of personalized diagnosis and treatment of each patient.

7.2.3.5 Genomic Medicine

According to Dr Bekiaranov, Dr Maria Schuld, Quantum Machine Learning Developer at Xanadu, a Toronto-based photonic quantum computing and advanced artificial intelligence start-up, was one of the first to generate "implementable, near-term" quantum machine learning algorithms. This new quantum classifier, meant to be used on genomic data, will determine whether a test sample derives from a disease or control group at a much faster speed than classical computers. For example, classical computers require 3 billion operations to categorize four building blocks of human DNA (i.e., A, G, C, T) whereas quantum algorithm needs only 32. All of our computing systems are able to store and manipulate information. Today's, conventional digital computers perform calculations using bits that can, like on-and-off switches, only exist in two states (0 and 1). Quantum computers, instead of manipulating individual bits, use quantum bits (qubits), which enables them to encode information as 0s, 1s, or both at the same time. This is called a superposition state and permits quantum computers to manipulate huge combinations of states at once. Quantum computers process a huge amount of data and also more quickly than traditional counterparts.

This kind of performance could greatly help certain sectors such as the health sector, especially genome sequencing, the process of determining the order of the four nucleotide bases: Adenine, Guanine, Cytosine, and Thymine that constitute a DNA molecule. The human genome is made up of over 3 billion of those bases. Studying the entire genome sequence will help us understand better how genes work together. Searchers are looking for a low-cost sequencer. Today, getting your full genome sequenced is still expensive and not affordable by everyone, at approximately $1,000. So, having a lower-cost and rapid sequencing method would enable to sequence more genomes. Today's methods involve dividing the DNA into small parts to search certain types of biomarkers and disease-related mutations. This brings two main consequences: the process is too time-consuming and is slowed by the manual operation; plus, genome sequencing needs a lot of storage and computational power. Classical computers are not powerful enough for this task. Quantum computers can be a solution to address this problem because they have more computational power and storage than classical computers. In contrast with the latter, quantum computers would be able to find the order of the DNA bases faster [29]. Furthermore, they would be much more accurate in their findings than classical computers. This opens vast possibilities, like giving a proper diagnosis to patients and even personalized medicine, in order to have a better recovery. Quantum computers will revolutionize treatment by taking into account different factors that we encounter in our everyday lives like our environment or even lifestyle. In addition, quantum computers will allow us to generate a database of genomes that could unveil unknown biomarkers and mutations furthering our understanding of how our DNA is encoded.

7.2.3.6 Data Storage, Transmission, and Security

This is a metaphor for superposition as it applies to quantum mechanics. One bit (the cat) can have several states at the same time and is, therefore, fundamentally different from the classical on/off or 0/1 representation in today's computer science, which is based on physical laws. Due to this possibility of superposition states, parallel computing operations can also be performed with quantum laws enforcement, which accelerates the time of complex calculations. Google announced a few months ago that they have managed to build a quantum computer with 53 (Q) bits, capable of handling computations much faster than current supercomputers can; it can solve a selected problem in 3 minutes instead of 10,000 years, for example.

QC definitely brings in changes in the encryption process. Most difficult and large time taking asymmetric algorithms with the factorial level of computations to crack by traditional computer will become crackable easily with the introduction of IT. MIT demonstrated this feature already. However, in the case of symmetric algorithms, quantum necessitates multiplication of key sizes. For example, large quantum computers run Grover's algorithm, which uses quantum concepts to survive attacks. They want key sizes to be doubled to make protection more secure. Though not yet introduced, quantum cybersecurity is the need of the near future.

Although there are possible risks, the security aspect of QC [30] yields many vigorous and absorbing factors for protecting vital and individual data. This can be more seen in machine learning enabled by QC and quantum random number generation applications.

7.2.3.7   Predictive Healthcare

Every stage of patient treatment can be well tracked by predictive analytics. Patients can be monitored continuously during phases such as detection, forecast, and cure. It also supports features like faraway patient observation and reducing bad events due to illness. It can also lead to a significant reduction in medical costs.

> **Detection:** Predictive analytics facilitate diagnosis of illness at the early stage leading to the possibility of early treatment and improving the chances of survival a lot. Patients can get treatment for their ailments at the starting stage of the disease itself.
>
> **Forecast:** Predictive analysis enables researchers to understand the physiological aspect of patients with serious illnesses. This prevents congestive heart failure in patients. It also minimizes the risk of readmission and recurrence of the problem. This feature facilitates having a good prediction of readmissions.
>
> **Cure:** Machine learning equipped with quantum mechanics principles leads to the determination of effective treatment course for chronical diseased patients. This will provide a very effective cure for the ailment of the patient.

The goal of predictive healthcare is to reduce healthcare costs while improving patient outcomes.

Predictive analytics is an advancing method of improving patient outcomes. By looking at data and outcomes of past patients, machine learning algorithms can be programmed to provide insight into methods of treatment that will work best for the current patients.

Additionally, predictive analytics can be used to identify warning signs before conditions become severe. With the COVID-19 pandemic being at the forefront of healthcare, researchers are putting resources into developing predictive analytic methods for combating the virus.

COVID-19 has been a unique struggle for the healthcare system since the severity of the virus can vary from person to person. In a recent study posted on *JAMA*, researchers highlighted risk factors associated with severe cases of COVID-19 by using machine learning algorithms and predictive analytics. The study concluded that demographic characteristics and comorbidities were among the highest risk factors.

Predictive analytic methods are also being used to determine the severity of COVID-19 from person to person. American Chemical Society's researchers recently developed a blood test that uses predictive analytics to project whether an individual will experience severe COVID-19 symptoms or not.

By using predictive analytics and a process called Attenuated Total Reflectance – Fourier Transform Infrared Spectroscopy (ATR-FTIR) [31], researchers determined that the best indication of whether a patient would experience severe COVID-19 symptoms is if the patient has diabetes.

Predictive analytic methods allow providers to determine individuals at risk for developing severe infections or chronic diseases.

By identifying those at risk, it provides medical professionals with an opportunity for early intervention and chronic disease prevention. With predictive analytics, providers can identify patients who are potentially at high risk for certain chronic conditions such as cancer, cardiovascular disease, diabetes, obesity, and kidney disease.

7.2.3.8 Hospital on Cloud

Quantum emulators, simulators, or other processors are provided as services by cloud enabling cloud-based quantum computing [32]. They embed huge computing power into it by harnessing through emulators or simulators that can execute work in a parallel manner. This infrastructure need not be acquired by each medical company. It can have it from the cloud as per its requirement. This gives rise to pay for what is used. This will be definitely less than the cost to have set up of its own.

Its working [33] is shown below:

1. Developers developed one tool named pyQuil in which quantum images can be interfaced with traditional systems. They are called quantum machine images (QMI). These images are actually analysed through virtual programming and processing units. Such environments are suitable to run quantum software applications.
2. Quantum Virtual Machines (QVM) perform the execution of such coded programs. They run quantum processors to perform that task and obtain its results. They also test the program and generate its waveform as a result.
3. Another setup called Quantum Processing Unit (QPU) runs the waveforms generated from quantum machine images. The configuration of qubits is done using these waveforms.
4. QPU prepares required information and then QMI executes it. The result is finally given back to the classical computer.

Some of the cloud-based QC providers are Microsoft Azure Quantum, IBM Q Experience, Amazon Braket, Google's Quantum Playground, Rigetti Forest, D-wave Leap, and Xanadu.

7.3 FUTURE SCOPE

The possible future scopes in a QC-based healthcare system are as below:

1. **Hospitals on Clouds:** They are voluminous in nature and also need temperatures about 15 millidegrees above absolute zero to operate. They require huge computation units those require more space naturally. Because of these features, they are hard to install at the required location rather than making them available over the cloud.

2. **Cognizance of Quantum:** It includes human brain like unit, language skills, decision-making ability, memorizing capability, and analytical abilities in task doing. Quantum probabilities play a major role in all these activities of cognizance of quantum. This aspect plays a vital role in healthcare in order to think like the human brain in tackling health problems and providing their solutions.

3. **Cryptography via Quantum:** It aims the development of encryption methods that are stronger than the traditional ones available at present. Quantum cryptography gives scope for the development of such methods to strengthen security further. They draw that strength from the quantum mechanics properties of particles. If anyone wants to grab encoded data, the quantum state captures that attempt. It shows that attempt by changing its state immediately. This facilitates noticing such an attempt. Then it can be prevented from taking place. The healthcare system with this option will provide for high security of medical data.

4. **Quantum Neural Network (QNN):** Neural networks can be extended to a new domain called Quantum-oriented Neural Networks. They have features of ordinary neural networks along with embedded quantum principles. This will provide an opportunity to develop more efficient algorithms. These algorithms help to extend networks, memory devices, and control systems automation for the quantum environment. This will enable for automation of the overall healthcare system.

7.4   CONCLUSIONS

The following conclusions can be drawn from this research article:

1. Improvements in all aspects of different stages of the healthcare system, namely, imaging, screening, prediction, curing, and monitoring health can be seen. Also, the new healthcare system is able to manage a huge number of patients in a very nice manner.
2. QC facilitates a more focused and precise analysis of any disease irrespective of its nature.
3. Medicines for diseases can be produced and marketed with supersonic speed.
4. Uncurable and hard diseases of the present can be handled and cured.
5. Medical data can be kept in an ultra-secure manner.

ACKNOWLEDGEMENT

I am thankful to Dr Shankar Kumar Ghosh for his constant support and for providing important suggestions which improved the quality of the work.

REFERENCES

[1] https://reports.valuates.com/market-reports/QYRE-Othe-0A163/global-digital-health

[2] https://pubmed.ncbi.nlm. nih.gov/ 1606330/

[3] https://www.healthcarefacilitiestoday.com/posts/ The-impact-of-medical-computers-on-patients-8931

[4] https://www.euro.who.int/en/health-topics/health-emergencies/coronavirus-covid-19/novel-coronavirus-2019-ncov - Pandemic Announcement Info.

[5] Naylor CD, Chantler C, Griffiths S. "Learning from SARS in Hong Kong and Toronto", *JAMA* 2004, 291, 2483–2487. DOI: 10.1001/jama.291.20.2483.

[6] Danquah LO, Hasham N, MacFarlane M, et al. "Use of a mobile application for Ebola contact tracing and monitoring in northern Sierra Leone: a proof-of-concept study", *BMC Infect Dis* 2020, 19, 1–12.

[7] https://www.nature.com/ articles/s41591-020-1011–4#Sec9

[8] Czeisler ME, Marynak K, Clarke KE, et al. "Delay or avoidance of medical care because of COVID-19-related concerns – United States", *MMWR Morb Mortal Wkly Rep* 2020, 69, 1250–1257. DOI: 10.15585/mmwr.mm6936a4.

[9] https://www.grandviewresearch.com/industry-analysis/digital-health-market

[10] https://www.dqindia.com/ digital-transformation-can-accelerate-growth-indian-health-care-landscape-avenues/

[11] https://www.controleng.com /articles/quantum-computing-used-to-discover-possible-covid-19-treatments/

[12] https://medicalfuturist.com/quantum-computing-in-healthcare/

[13] https://bcmj.org/ blog/quantum-computing-health-care

[14] https://www.drishtiias.com/daily-updates/daily-news-editorials/quantum-technology

[15] https://www.globenewswire.com/en/news-release/2021/01/04/2152292/28124/en/Global-Smart-Healthcare-Market-is-Forecast-to-Grow-by-224-86-Billion-during-2020-2024-Progressing-at-a-CAGR-of-24.html

[16] Google AI Quantum and Collaborators, "Hartree-Fock on a superconducting qubit quantum computer". *Science* 2020, 369, 1084.

[17] Sajjan M, Sureshbabu SH, Kais S, "Quantum machine-learning for eigenstate filtration in two-dimensional materials", *J Am Chem Soc* 2021, 143, 18426–45.

[18] Deutsch D, "Quantum theory, the Church-Turing principle and the universal quantum computer", *Proc R Soc Lond A Math Phys Sci* 1985, 400, 97–117.

[19] Annunziata A. IBM Quantum Summit 2020: "Exploring the promise of quantum computing for industry", 2020.

[20] Adrian C. IBM promises 1000-qubit quantum computer—a milestone—by 2023, Science, 2020.

[21] Bharti K, Cervera-Lierta A, Kyaw TH, et al. "Noisy intermediate-scale quantum (NISQ) algorithms", arXiv:2101.08448, 2020.

[22] Kottmann JS, Schleich P, Tamayo-Mendoza T, Aspuru-Guzik A, "Reducing qubit requirements while maintaining numerical precision for the variational quantum eigensolver: a basis-set-free approach", arXiv:2008.02819, 2020.

[23] Chen SY, Wei TC, Zhang C, Yu H, Yoo S, "Quantum convolutional neural networks for high energy physics data analysis", arXiv:2012.12177, 2020.

[24] Malviya R, Sundram S, "Exploring potential of quantum computing in creating smart healthcare", 2021. DOI: 10.2174/1874196702109010049.

[25] Rasool RU, Ahmad HF, Rafique W, Qayyum A, Qadir J, "Quantum computing for healthcare: a review", 2020.

[26] Chen SY, Huang CM, Hsing CW, Kao YJ, "An end-to-end trainable hybrid classical-quantum classifier", *Mach Learn: Sci Technol* 2021, 2, 045021.

[27] Drug Discovery Online. "Researchers use NSF convergence accelerator to shorten drug discovery timeline", 2020. Accessed November 2020.

[28] Fadziso T, Edge computing and quantum computing to find statistics of pandemic", *Eng Int* 2018, 6, 143–154.

[29] https://projectqsydney.com/ transforming-drug-development-a-critical-role-for-quantum-computing/

[30] For a brief on quantum cryptography, see for instance ETSI White Paper No. 8 "Quantum Safe Cryptography and Security An introduction, benefits, enablers and challenges" June 2015.

[31] Compton K, Edwards C, Komenda P, Bose A, Milad J. Human factors testing of the quanta SC+: demonstrating ease of use with minimal upfront training in health care practitioners and patients, *Nephrology Dialysis Transplantation* 2020, 35, gfaa142. P1492. DOI: 10.1093/ndt/gfaa142.P1492

[32] https://research.aimultiple.com/quantum-computing-cloud

[33] https://searchcloudcomputing.techtarget.com/tip/The-future-of-quantum-computing-in-the-cloud

8 An Overview of Future Applications of Quantum Computing

Taskeen Zaidi and Bijjahalli
Sadanandamurthy Sushma
Jain (Deemed to be University)

CONTENTS

DOI: 10.1201/9781003250357-8

8.1 INTRODUCTION

Quantum computing provides a new approach to computing. The recent trends of quantum computing are mainly based on the phenomenon of quantum mechanics to solve many complicated problems. It is a combination of multiple scientific domains such as mathematics (algorithm and arithmetic), physics, computer science, statistics and information theory. The quantum computers provide high computational speed, optimize energy consumption and are efficient in handling high-performance tasks. In quantum computing, subatomic high energy particles like atoms, electrons, photons and other transition ions are used and all the subatomic particles possess spins and transition states. These particles can be superposed in order to provide more new combinations. Parallel tasks can be executed on quantum computers with efficient memory utilization. The performance of quantum computers is better than that of classical computers. The theory of quantum computing is applied in many existing algorithms in modelling, simulation and complexity theory to solve mathematical problems. Some of the quantum computing strategies were used to design next-generation processors and software for computing. Quantum computing will bring drastic changes in information technology. The superposition principle provides the ability for the quantum computer to be in multiple states. The quantum system exists in 0 or 1 state, and superposition is the characteristic of the quantum computer. The entanglement principle creates and manipulates quantum entities. The quantum computer is able to create quantum operations using quantum algorithms on qubits. Quantum computing brings a revolution in computations by solving intractable and complex problems in an effective manner. The quantum computer has the capability to provide millions of times better performance than present-day computers. The advancement in information technology due to quantum computing is one of the potential topics of research across the world. A brief history of quantum computing, qubits, operator and measurement, the hardware applications, quantum computing methods and applications of quantum computing in various areas is well explained [1]. The authors explained the quantum computing approach; also, the problem solved using quantum computing approach is well explained [2,3]. The quantum computer operations, Born rule, Deutsch's problem, Simon problem, Rivest Shamir Adleman encryption, quantum error correction and protocols used in quantum cryptography are well explained [4]. The fundamental concept of quantum computing, communication protocols, quantum metrology, quantum correlations, quantum entanglement and advanced concept of quantum computing is well explained [5,6]. The theory of quantum computing including the interdisciplinary approach to quantum computing in fields like engineering, computer science, medicine and life sciences as well as tools and techniques used to solve complex computations were discussed [7]. The fundamental optical principles like diffraction, refraction and reflection were explained. The authors also discussed Dirac's notation and quantum entanglement [8]. The emerging area of quantum technology is covered in books like quantum cryptography, quantum teleportation and advancement in quantum computing with theory and experiments [9,10]. The theory of classical linear error correcting codes, properties of quantum systems, quantum stabilizer codes, fault tolerant quantum computing and fault tolerant protocols were discussed [11].

The author implies single-photon two-qubit quantum logic to implement a photonic attack against Bennett-Brassard's 1984 protocol [12]. As per a study, the researchers from Google identified that quantum computer has the ability to perform tasks which require high-performance computing [13]. A quantum algorithm is proposed to estimate the performance of existing quantum algorithm for linear system of equations [14]. The quantum computer uses the properties of quantum mechanics for computations. It maintains coherence and is able to detect and correct errors. The quantum computer is based on semiconductor technology which operates at zero temperature to minimize the de-coherence [15]. In quantum computing, the information is represented in qubits. The qubit states are written as 0 and 1. It also follows the phenomenon of superposition [16]. The essential elements of quantum theory, a comparison between classical and quantum computing, superposition, entanglement, quantum programming and quantum storage management were well discussed [17]. Google Artificial Intelligence quantum team has conducted chemical simulations on a quantum computer. The potential benefits of a quantum computer are discussed [18].

8.2 QUANTUM COMPUTER FRAMEWORK

The framework of a quantum computer is classified into five layers, and each layer is represented as part of the computer, as shown in Figure 8.1.

i. **Application Layer:** It represents the interface, operating system, quantum computer, etc. It is hardware dependent and performs quantum algorithms.
ii. **Classical Layer:** This layer optimizes and compiles quantum algorithm into microinstructions.
iii. **Digital Layer:** This layer interprets microinstruction into signals. It merges the quantum outcomes into final results.
iv. **Analog Layer:** This layer produces high-voltage signals and sends them to the below layer and then qubits operations can be performed.
v. **Quantum Layer:** It integrates digital and analog processor layers into the same chip. Error is corrected by this layer and computer performance is also evaluated.

The quantum processing unit comprises of platforms such as digital processing layer, analog processing layer and quantum processing layer.

8.3 EXPERIMENTAL PERSPECTIVE OF QUANTUM COMPUTING

The simple model of a quantum computer may be faster, cheaper and more efficient. The quantum computers may replace the classical computers in future. Some of the attempts to develop an efficient quantum computer are listed below:

i. **Heteropolymer-based Quantum Computer:** Heteropolymer-based quantum computer was designed in 1988 and further improved by Lloyd (1993). The memory cells were composed of a linear array of atoms and the atom was pumped into an excited state and laser pulses were used for transmission

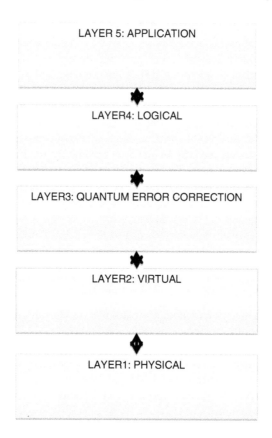

FIGURE 8.1 Architecture of quantum computing

of instruction. The computation was done on the shape and duration of the pulse on selected atoms.

ii. **Ion Trap Quantum Computer:** Cirac and Zoller proposed the ion trapped quantum computers in 1995. The qubits were used for computation.

iii. Turchetta implemented Quantum Electrodynamics (QED), which has a QED cavity filled with Caesium atoms. The quantum computer followed the principle of superposition and entanglement.

iv. **Quantum Dot Technology:** The quantum dots are semiconductors with dimension between 10 and 100 nm. The quantum dots are best for implementing quantum logic gates. An array of quantum dots connects the nearest neighbour through the split-gate technique.

8.4   QUANTUM COMPUTING ALGORITHMS

A quantum computer works on a quantum algorithm. The algorithm contains at least one unique quantum either superposition or entanglement for execution on quantum computers. The quantum circuit and quantum gate are also required. The quantum gate can be proposed on the number of qubits. There may be single qubit or multiqubit gate. The quantum algorithms are reversible and undo all the operations by forward traversing. A few examples of quantum algorithms are mainly Shor's and Grover's algorithms. Shor's algorithm factorized large numbers faster than the classical algorithm, whereas Grover's algorithm is used for searching unordered list faster than the classical algorithm. Various quantum algorithms like algebraic and number theoretic algorithms, oracular algorithms, simulation algorithms and optimization algorithms are well discussed [19].

8.5 QUANTUM COMPUTING APPLICATIONS

8.5.1 QUANTUM COMPUTING HEALTHCARE APPROACHES

The quantum computing application may be helpful in drug discovery and process optimization. The classical computer is not able to solve optimization problems due to a number of variables and computations; so, the quantum computer may be one solution to perform different options at the same time. IBM installed its quantum computer first time on an offsite private business. These computers implemented the applications related to genomics, infectious disease and public health. Then, later on, National Player Anthem integrated with IBM's network hub for developing a more reliable and personalized treatment for the prediction of health conditions. The quantum computing applications include the diagnosis of diseases, advancing precision medicine and optimizing pricing and risk analysis.

The molecules comparison can be done for drug design and discovery on quantum computers as classical computers are of limited size but quantum computers have the ability to compare large molecules.

8.5.2 DIAGNOSTIC ASSISTANCE

The medical imaging techniques like Computer Tomograpgy, Magnetic Resonance Imaging and X-ray scans become popular for diagnosing diseases. But many images obtained by imaging techniques are impacted by noise, poor quality, poor resolution and low replaceability. Quantum computing has improved the analysis of medical images using edge detection and image matching. The biomarkers may be helpful in genomics, transoptomics, proteomics and metabolomics. Quantum computing may be helpful in discovering biomarkers even for individuals. Quantum computing diagnoses may monitor and analyse the health of an individual.

It would be easy to diagnose elaborate diagnostic procedures for determining the procedure type and it will reduce the costs also. Quantum computing can also boost AI capability for the diagnosis of diseases at an early stage. The high-resolution MRI can be helpful in the screening of diseases.

8.5.3 PRECISION MEDICINE

Precision medicine is also helpful in the prevention and treatment of the individual. Many existing therapies fail to detect the adverse effect of drug reactions. Early treatment and preventive intervention may reduce the risk of diseases but only risk analysis is not sufficient, it is important to detect how much medical intervention is needed for an individual. The quantum-enhanced machine learning is able to support and predict the effectiveness of drugs. Precision medicine is able to provide the best medical relevant data and can be efficient for checking health status. Quantum computing may be able to accumulate the risk related to the umbrella approach.

Drug development requires lots of trials and a number of errors are detected at the preliminary stage. This process is risky and expensive too. Quantum computing can be a game changer and it can analyse and perform computation for the drug development process. It also saves time and money. Classical computing is not helpful in predicting the reaction of a drug on the human body as every individual has their own genetic composition.

8.5.4 PRICING

The health insurance premium calculation is a complex process. There are a number of factors in a health plan for calculating the premium. Some of the factors are complex interdependencies, population health levels and disease risks, treatment, costs and risk exposure. Quantum computing may be helpful to optimize pricing risk. It may be a better approach to assess the risk a patient has for a given medical condition. Quantum computing may be helpful in fraud detection. The quantum computing algorithms enable superior classification and pattern detection to eliminate fraud during medical claims. The quantum computing algorithms may be helpful in improving pricing computations. This could reduce the premium.

8.5.5 QUANTUM TECHNOLOGY

Quantum computing is also applied in other scientific domains such as quantum information science, quantum communication, quantum physics and quantum metrology. All these technologies transform the quantum system. Quantum science applied the encoding of information in a quantum system. It used all the statistics of quantum mechanics and its limitations. Quantum mechanics is the core of other applications like quantum computing, communications, networking and metrology.

Quantum networking includes the exchange of information between quantum computers. Quantum cryptography includes quantum communication for securing the information exchange during data communication. The quantum entanglement states are distributed in quantum networks. The quantum network distributes the part of an entanglement state into various processors and acts independently. The quantum repeaters were used for quantum error correction and for repairing the signal without quantum information.

Quantum metrology includes the development of a quantum system that measures physical properties like electromagnetic fields, temperature, etc., for predicting more accurate than other classical systems.

Quantum computing utilizes quantum properties like superposition, entanglement and interference for computations. The qubits manipulate the algorithm to achieve the result with high quality and accuracy.

8.5.6   QUANTUM CRYPTOGRAPHY

The security of online e-commerce platforms uses encryption and mathematical algorithms which are not easily broken like factoring numbers into prime numbers. Modern algorithms like Advanced Encryption Standard (AES), Elliptic Curve Digital Signature Algorithm (ECDSA) cannot break using high computing power, but quantum computers are easily doing this in optimum time. The implementation of a cryptographic entangling probe is discussed [20].

8.5.7   ARTIFICIAL INTELLIGENCE

The AI technique applies Supervised, Unsupervised and Reinforcement learning approaches to large and complex datasets.

Quantum computing can perform training models within exponential time with higher accuracy. The generative models generate output as image, audio, etc., and the quantum computers improve their quality and accuracy.

8.5.8   WEATHER FORECASTING

Perhaps, it is not possible to obtain 100% accurate data and exact predictions about weather situations and to issue early warnings about bad weather conditions. A very high computational power is required to predict the weather. The classical computers are not able to detect it. Quantum computers can be useful in weather forecasting as quantum computers are able to analyse large data resources in optimum time. The quantum computer can be helpful in efficient weather forecasting and also helpful in saving properties and lives. The quantum machine learning approach can be helpful in pattern recognition which will be helpful in weather changes and abnormal conditions.

8.5.9   FINANCIAL MANAGEMENT

Quantum computing plays an important role in financial modelling and risk analysis during payment transactions. The strategies can be optimized for reducing risk. The quantum systems are able to select an optimized portfolio in order to maximize returns. The quantum systems are able to detect fraudulent transactions. Quantum computers can be helpful in predicting investments related to the financial industry. Currently, all the predictions were done on classical computers using Monte Carlo simulation and it is a very time-consuming operation.

The quantum drone is also an emerging technology. The scientists and researchers were designing drones to transmit particles of light. The photons can be sent to the two different locations covering a few kilometres as per the study published by Nanjing university researchers published in physical review letters. The scientists are aiming to create global quantum internet that enables ultra-secure communication using encryption algorithms. The quantum internet can connect distant quantum computers also for information interchange. The quantum network is created using two drones to transmit photons.

The military technology requires greater caution for deployment in the battle-ground. The quantum sensors can be used in place of classical sensors. New hardware and systems are required to deploy with the current system. The quantum computer can be used in long term. Quantum technology can be useful in warfare. The integra-tion of the quantum system in military applications is practically more challenging. The deployment requires portability, sensitivity, speed, performance, security, cost and also a high-performance laboratory.

8.5.10 LOGISTICS OPERATION

Logistics scheduling requires a lot of optimization and monitoring during supply chain management. The models require traffic management randomly. The classical is handling the logistics operation but it is a little bit complex and time-consuming; the quantum technology can be helpful in optimizing the supply chain management.

8.5.11 QUANTUM KEY DISTRIBUTION

The quantum key distribution is referred to as quantum cryptography which securely generates a cryptographic key that can be exchanged without causing any malicious attack. The key is generated by the random spin of the particle and the key is used in conjunction with conventional symmetric algorithms.

8.5.12 QUANTUM COMPUTING IN MACHINE LEARNING AND BIG DATA

Machine learning organizes the items into large data sets, and quantum computing is useful in classifying the data. The Quantum Random Number Generator (QRNG) generates random numbers using quantum physics principles which offer maximum entropy. The QRNG can be useful in areas like AI, Machine Learning, Internet of things (IoT) and Soft computing for communications.

In the current world, data is generated at a faster rate. Data handling and processing require new technologies and algorithms. Traditional and classical computing is not able to process a massive amount of data. According to Moore's prediction, we require more computational power to speed up the task execution; and in the current scenario, quantum computers can be a solution. We need to solve every complex problem at a faster rate and size within the optimum time. The processing power is also required to process a large amount of data; so, quantum computing will be a feasible solution.

8.5.13 QUANTUM COMPUTING IN CLOUD COMPUTING

The integration and assimilation of cloud and quantum computing will be the future of next-generation computing. The quantum computer is used to create a super-fast computer using quantum physics principles, and cloud computing is used to provide various services and resources to the user on demand. The quantum emulators can be accessed through cloud-based quantum computing. The processing speed of a quantum computer can be accessed by users over the internet through quantum pro-cessing in the cloud environment. The issues of security, back-up and computational speed-up were resolved using quantum approaches. The Public Key Infrastructure (PKI) and Kerberos are used for exchanging keys using the certificate on a public platform for validation purposes. The quantum cloud can be helpful for securing, storing and processing data efficiently.

8.5.14   QUANTUM COMPUTING IN ROBOTICS

The quantum properties like entanglement can be helpful in analysing visual information. Quantum Image Processing (QIMP) is an emerging area in analysing visual information. The AI approaches can be used in robotics for modelling and analysing a system through quantum algorithms. The task planning and specific movement planning problems can be solved using quantum approaches integrating a dynamic programming paradigm. The fault detection and isolation can be easier using quantum computing.

8.5.15   QUANTUM COMPUTING IN MATERIAL SCIENCE

Quantum computing can bring advancement in the field of material science for creating alternatives towards emerging greener technologies. An application like high-temperature superconductors can be helpful in energy transmission with minimum loss. The new material can be identified with high superconductivity properties using quantum technology.

8.6   CONDITIONS FOR QUANTUM COMPUTING

The basic conditions for quantum computations are:

1. Representation of quantum information
2. Unitary transformation
3. Representation of input
4. Measurement of output

Quantum computing is based on the particle transformation of quantum states. The quantum bits labelled the pairs of states and their realization. The states should be finite. The unrecorded errors in transformation cause de-coherence and also lead to errors in quantum information represented in the form of quantum state. The closed quantum systems are unitary; but for performing quantum computation, it is important to control Hamiltonian to select unitary transformations. The useful computation can be performed on desired input. The input state preparation is a difficult problem for the physical system. The fidelity and entropy should be considered for input state preparation. The output result measurement is difficult due to inefficient photons, noise, etc.

8.6.1   ADVANTAGES OF QUANTUM COMPUTING

1. Quantum computer is generally able to solve complex tasks such as mathematical problems that were not solved by traditional computers.

2. It provides computational power for processing a large amount of data and it reduces power consumption also and is even much faster (thousands of times) than classical computers. It can solve complex problems without being overheated.

3. It solves optimization problems easily and can compute 1 trillion moves in chess per second.

4. Quantum computers are capable of creating secure encryption techniques and can be used in various organizations and financial organizations.

8.6.2   Challenges in Quantum Computing

The quantum computers might suffer from coherence. The challenge is to handle inexpedient deviations and noise in quantum computers. The qubits may entangle as qubits are neither completely binary nor digital. It has analysis properties also, but there is a requirement for an algorithm that can run on a noisy gate-based quantum computer to reduce the induced noise. The quantum error correction algorithm may be one solution but it shows complexity in handling a large number of qubits and performing arithmetic operations. The data inputs require abundant time to generate an input quantum state. The processing speed will also become low for computation. The debug quantum system may be helpful to check unknown issues. New debugging strategies are required for the development. The quantum computers are still not fully implemented as it is only invented in parts. It gets affected by noise, faults and failures as subatomic particles are affected by any kind of vibrations.

The errors were introduced and quantum computers suffer from the gradual de-coherence and the rate of de-coherence depends on the technologies used for implementing quantum computing. The de-coherence errors may not impact the quantum computer but the errors gradually increase and complexity also increases over time. The de-coherence errors are random with bounded magnitude and to reduce to reduce the quantum de-coherence and extra qubits were proposed by the Shor and Steame and they also added the elimination of the errors in superposition. A fault tolerant algorithm on quantum networks was proposed by Shor and Kitaev.

The quantum processors are very difficult to manage and handle due to their instability and testing also requires a lot of effort. The production and maintenance cost of quantum computers is also very high.

The current gate-based quantum computers have less qubits, but for performing complex computations qubits requirement is more. The scaling issue will create difficulty in connecting and operating qubit. The qubits require retention time to perform complex algorithms; and in the current qubits type, it is a big challenge to extend the coherence time. The error correction code implementation is a big challenge. It is difficult to perform reliable and fault tolerance operations on current quantum computers.

8.7   FUTURE OF QUANTUM COMPUTING

1. **Optimization Problems:** By using quantum annealing, the optimization problems can be solved efficiently and faster.

2. **Machine Learning/Big Data:** Quantum computers are faster in training and testing data sets. So, quantum computing may be helpful in machine learning and deep learning for efficiently training and testing models.
3. **Simulation:** It helps to solve complex simulation problems.
4. **Material Science and Chemistry:** The quantum computing will be helpful in solving the complex interactions of atomic structures. The quantum solution will offer an efficient way to solve these interactions.

8.8   DISCUSSION

This chapter provided an overview of quantum computers. The architecture of quantum computing is discussed in detail as well as the experimental perspective of quantum computing is explained. The quantum algorithms related to different areas like healthcare, diagnostic assistance, AI, precision medicine, quantum technology, quantum cryptography and weather forecasting are discussed. The condition of quantum computing, the advantages and disadvantages of quantum computing as well as the challenges and future direction of quantum computing are well explained.

REFERENCES

1. J. D. Hidary, Quantum Computing Methods. In: *Quantum Computing: An Applied Approach*, Springer, 2019. Springer, Cham. https://doi.org/10.1007/978-3-030-23922-0_9.
2. E. R. Johnston, N. Harrigan, M. Gimeno-Segovia. *Programming Quantum Computers*: essential algorithms and code samples. O'Reilly Media, 2019.
3. N. S. Yanofsky and M. A. Mannucci, Quantum computing for computer scientists, *Contemporary Physics*, 50:6, 668–669, 2009. DOI: 10.1080/00107510902986702.
4. N. D. Mermin, *Quantum Computer Science: An Introduction*, Cambridge University Press, Cambridge, 2007. doi:10.1017/CBO9780511813870.
5. G. Leuchs and D. Bruss, *Quantum Information, 2 Volume Set*, 2nd Edition, Wiley-VCH, 2019.
6. V. Silva, *Practical Quantum Computing for Developers: Programming Quantum Rigs in the Cloud using Python, Quantum Assembly Language and IBM QExperience*, Apress, 2018.
7. M. Ying, *Foundations of Quantum Programming*, Morgan Kaufmann, 2016.
8. F. J. Duarte, *Quantum Optics for Engineers*, CRC Press, 2017.
9. D. Bouwmeester, A. K. Ekert, and A. Zeilinger, *The Physics of Quantum Information: Quantum Cryptography, Quantum Teleportation, Quantum Computation*, Springer Verlag, Berlin, German, 2007.
10. W. P. Schleich and H. Walther, *Elements of Quantum Information*, Wiley-VCH, 2007
11. G. Frank, *Quantum Error Correction and Fault Tolerant Quantum Computing*, CRC Press Inc, 2008.
12. T. Kim, I. S. genannt Wersborg, F. Wong, and J. H. Shapiro, Complete physical simulation of the entangling-probe attack on the Bennett-Brassard 1984 protocol, *Phys. Rev. A* 75, 042327, 2007.
13. M. Giles, Google researchers have reportedly achieved "quantum supremacy". Retrieved December 07, 2020, from https://www.technologyreview.com/2019/09/20/132923/google-researchers-have-reportedly-achieved-quantum-supremacy/
14. A., Harrow, A., Hassidim, and S. Lloyd, Quantum algorithm for linear systems of equations. 2009. Retrieved December 06, 2020, from https://journals.aps.org/prl/abstract/10.1103/PhysRevLett.103.150502

15. W. C. Holton, Quantum computer (1107908068 837841751 Encyclopædia Britannica, Ed.). Retrieved December 07, 2020, from https://www.britannica.com/technology/quantum-computer

16. QuTech Academy, What is a qubit?Retrieved December 06, 2020, from https://www.quantum-inspire.com/kbase/what-is-a-qubit/

17. M., Rouse and B. Pawliw, What is quantum computing? 2020, Retrieved December 06, 2020, from https://whatis.techtarget.com/definition/quantum-computing

18. B. Yirka, Google conducts largest chemical simulation on a quantum computer to date. 2020. Retrieved December 07, 2020, from https://phys.org/news/2020-08-google-largest-chemical-simulation-quantum.htm

19. Quantum Algorithm Zoo, (Compiled list of Quantum algorithms), http://quantumalgorithmzoo.org/

20. E. E.Howard, Quantum-cryptographic entangling probe, *Phys. Rev. A* 71, 042312, 2005.

9 Authentication and Authorization for Electronic Health Records Using Merkle Tree – Attribute-Based Encryption

Mahalingam P.R.
InApp Information Technologies Pvt. Ltd.

CONTENTS

DOI: 10.1201/9781003250357-9

9.1 INTRODUCTION

The scope of automation in healthcare has progressed rapidly over the last decade owing to advances in predictive analysis, availability of large pools of data, availability of better computing power, and advances in cyber security. Data is increasingly being stored in the digital form, with hospitals moving to online patient management portals. This has resulted in the reduction of paper-based patient records and delays in data access, leading to better treatment opportunities. The world is witnessing what is called "Healthcare 4.0" [15,24], which is the fourth revolution in healthcare, powered by IT and automation. The rest of this chapter deals with how privacy is handled in the era of automation, and a solution that may be used to ensure the authenticity and validity of patient data access.

9.1.1 HEALTHCARE 4.0

The domain of healthcare has been looked upon as similar to the conventional industry in terms of different levels of development and advances happening there. While the Internet of Things and automation brought about Industry 4.0, the theme of "smart and connected healthcare" introduced the concept of Healthcare 4.0. The latest advancement is targeted at improving the socio-economic impact of healthcare by incorporating aspects like process improvement, technology-assisted process, etc., leading to optimized delivery of healthcare services. Healthcare 4.0 has been widely envisioned as a consequence of Industry 4.0, leading to smart, IT-enabled, and sustainable processes.

The growth from Healthcare 1.0 to 4.0 has seen patient care expand from a linear flow through a number of healthcare professionals to a network of healthcare professionals and caregivers, with ample automation and communication. This has resulted in big data streams with a variety of formats, characteristics, and quality. In addition, the involvement of patients has increased dramatically, with they being a part of initiatives to manage the level of access to their data [4].

9.1.2 ELECTRONIC HEALTH RECORDS

Electronic Health Record (EHR) refers to a digitized form of patient data that enables tracking of each person's treatment, diagnosis, procedures, prescriptions, and decision insights [12]. This concept was initiated as part of Healthcare 3.0, and 4.0 resulted in adding connected care, personalized medicine, and Artificial Intelligence-based predictions on the top of this collected data. The advent of EHRs has enabled insights to be generated not only based on a person's data but a pool of patients whose symptoms and treatment can be compared with the person. This facilitates the use of classifications, clustering, predictions, stratification, proactive care, timely interventions, early detection, and a closed-loop healthcare system.

As per Hoerbst and Ammenwerth, a typical EHR needs to consider the following:

- Functionality
- Data security

- Consent management
- Usability
- Interoperability
- Reliability
- Privacy and data protection
- Maintainability
- Performance/efficiency
- Portability

The features required for EHR can be categorized into two types:

1. **Architecture-based:** This incudes functionality, usability, interoperability, reliability, maintainability, performance/efficiency, and portability
2. **Security-based:** This includes data security, consent management, privacy, and data protection

Architecture-based requirements are managed by a proper choice of storage models, databases, and locations. For example, storage on the cloud may give good reliability and availability for the data. Choice of relational database and document-based database, which is based on the actual data itself, will determine how interoperable, maintainable, and efficient the access will be. The use of good data access methods will ensure EHRs are functional, usable, and portable.

At the same time, security-based concerns have to be addressed in a different dimension itself. Since EHRs have to be easily accessible during an emergency, they have to be stored on the internet front. Data security has to be ensured by using proper encryption methods, both at rest and in motion. Algorithms like Advanced Encryption Standard can maintain encryption at rest, while the use of Secure Sockets Layer/Transport Layer Security can ensure encryption in motion. Other security-based concerns have to be addressed using authentication and access control methods. They deal with the privacy of the stored data.

9.1.3 PRIVACY OF PATIENT DATA

Patient data contains sensitive information like Personally Identifiable Information (PII), disease data, diagnostic data, etc. From the patient's point of view, there should be proper dissociation between PII and diagnostics [2,6]. This ensures that even if the data gets leaked for some reason, there will be no way to trace it back to the patient. But the disadvantage of this approach is that the data becomes useless for healthcare workers as soon as the PII is dissociated. Hence, the ideal approach is that the data is loosely coupled with PIIs, and both are kept secure behind an effective authentication and authorization mechanism [18]. Methods like blockchain [6,16] are popularly used here.

9.2 ATTRIBUTE-BASED ENCRYPTION

Attribute-Based Encryption (ABE) is a popular method for managing the privacy of data and is widely adopted with EHRs. ABE implements a form of RBAC (Role-Based

Access Control) [13,22] that grants permissions to roles, instead of users. If we consider EHRs, it means that a patient may enable access to their records exclusively for cardiologists, rather than a single person. The person who tries to access the record will have to establish their identity as a cardiologist to gain entry. A typical RBAC entry for an EHR will look similar to the following.

Role:= (Cardiologist at wHealthCenter)
Allowed Action:= (Read OR Write Access)
Target:= (MR Patient 2342, Section: Laboratory Results)

This can be implemented using an AND-OR graph that needs to evaluate to true in order to gain access. ABE is implemented in one of two ways – Key policy ABE and Ciphertext policy ABE.

9.2.1 KEY POLICY ATTRIBUTE-BASED ENCRYPTION

In KP-ABE (Key Policy Attribute-Based Encryption), the communication is encrypted using a set of attributes, and private keys are associated with specifications on what type of ciphertexts the user will be allowed to decrypt [1,25,27]. Hence, each permitted user will be given their version of the key, with all necessary attributes. In short, user keys are associated with the roles themselves. This means that the source of data won't have any control over who has access to the data, since anyone with a valid role-based key would be able to decrypt and use the contents.

9.2.2 CIPHERTEXT POLICY ATTRIBUTE-BASED ENCRYPTION

In CP-ABE (Ciphertext Policy Attribute-Based Encryption), the key may remain common, but the access is restricted based on attributes tagged along with the ciphertext [5,10,26]. This corresponds to an RBAC where the privilege of decrypting and viewing information is vested upon only those individuals who satisfy the attributes tagged along with the ciphertext. On comparing with KP-ABE, it can be noted that CP-ABE associates the decryption privilege with users rather than keys. The main aim is collusion resistance, which means that unless at least one user is able to successfully decrypt a text, an entire team of cooperating users cannot gain access. In other words, users should not be able to simply combine their attributes to gain entry [17].

9.3 MERKLE TREES

Merkle trees give a convenient and effective method for the summarization and verification of large pools of data. The tree is based on a series of hashes [9,11] generated over a tree data structure. In simple words, the value of each node is the hash of concatenated hashes of all its children. It can be easily illustrated by considering the layout in Figure 9.1.

In Figure 9.1, the operation H1H2 simply means the concatenation of hash values H1 and H2. This layout ensures that the input data volume is very flexible. If

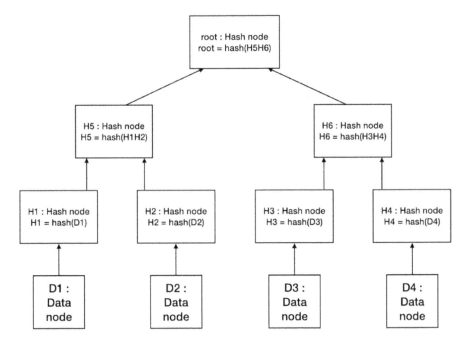

FIGURE 9.1   Basic working of Merkle trees

there are five data nodes, the extra node may be attached to either H5 or H6 as a third child, and the corresponding hash will be the concatenation of all three children. Since hash functions always generate a constant bit-width for inputs of variable sizes, adding any number of children to a node won't affect how the tree works. When Merkle trees are theoretically discussed, it is always preferred to keep the number of children to two. Hence, a fifth data block is best processed as in Figure 9.2.

Merkle trees take advantage of the fact that hashes are one-way functions. Hence, given a hash, it is virtually impossible to identify what value generated this hash. So, even if a Merkle root is available, it is not possible to identify the constituent data blocks [23]. In addition, any minor change in the input will result in widespread changes in its resulting hash. Hence verification is easy, making it suitable for a number of applications [3,7,14]. For all further consideration, Secure Hash Algorithm-256 is taken as the hash function to be used in the Merkle tree.

9.3.1   Role of Merkle Trees in Blockchain

Merkle trees rose in popularity with the introduction of blockchain. Merkle tree is used in blockchain to keep track of transactions within the block. Each block in the chain may consist of multiple transactions together. In that case, each is transaction placed in a dedicated block, and the Merkle tree is built. The Merkle root is then copied into the block for verification. When a block is required to be verified, the transactions are taken in the exact same way, and the Merkle tree is regenerated. If

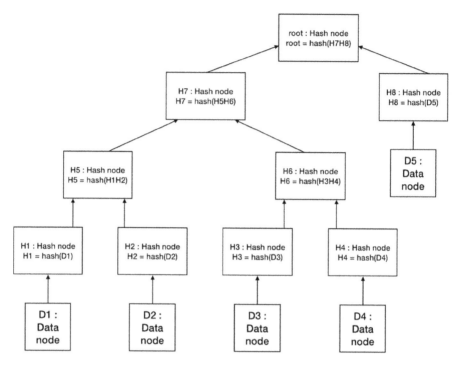

FIGURE 9.2   Preferred mode of operation with odd number of data blocks

the root matches the value recorded in the block, transactions can be said to be verified [19,21]. Let's consider it with an example, which is illustrated below.

Assume a block contains four transactions {{from: user1, to: user3, amount:145}, {from: user1, to: user2, amount:23}, {from: user2, to: user4, amount:98}, {from: user3, to: user2, amount:11}}

The Merkle tree as per Figure 9.1 will be built as follows.

D1: {from: user1, to: user3, amount:145}
D2: {from: user1, to: user2, amount:23}
D3: {from: user2, to: user4, amount:98}
D4: {from: user3, to: user2, amount:11}
H1 = b6ea9671ae98d5d58e41b0f0dc832ba48bd60e55dfb4e5449e-
        bff09301936bf5
H2 = 0d95ed56a0e77bd43e45436d1180796b29bb2dcaa0501b3e9dc3ff1fde-
        5dae7e
H3 = ec478b259223511db609b70ca897187713979a53b9be9bec6c3a1115f
        0359e36
H4 = d5d47b3762e2b3c27a877d5179ea478906b4c9c874f321315175e98c
        713392a8
H5 = acf5c13ca5a61d27ca513df9252909b3e17bbf0bee974872cdb4805d782f
        76b9

H6 = bab2bc39c00fc208cfe2319bb97eeef068c01a256aeca3fc5a24ba28fb
600411

root = 65412d867333c9da5c8070ecbb26d5ea7e57c659212757657a5540d-
4cf01e6ac

During verification, assume one of the entries got changed. Assume D4 got changed to {from: user3, to: user2, amount:110}. In that case, the change propagates as follows.

H4 = 40456c473a3f07269bdedad0699449a0a34add317cd9895d4045b-
f4c8b770728

H6 = 6684a274182e2d413fc79d5c99ee289d0681d699a14cd880c05d1578e-
c453eeb

root = 24ae733a66d392b29d31fdb52e06cea67f7b5499c5a64cf14188fc-
898cfa72e2

It can be seen that the root has changed to a good extent compared to the original value. When verifying in a blockchain, the intermediate values may not be available. Hence, tracing the error will be difficult. But it is possible to identify the presence as observed here.

9.3.2 String Commitment by Merkle Trees

String commitment is one of the most important properties of Merkle trees. While this is not a feature exclusive to these trees, the peculiar architecture of Merkle trees aids in performing string commitment. The string commitment scheme allows a sender to send a short commitment to a string s under any finite alphabet Σ.

- In the commitment step, the sender sends the root of the hash tree.
- If the sender is later asked to reveal the ith symbol in s, the sender sends the value of the ith leaf in the tree, as well as the value of every node v along the root-to-leaf path for s_i, and the sibling of each such node v. This is called the *authentication information* for s_i.
- The receiver checks that the hash of every two siblings sent equals the claimed value of their parent.

For each index *her*, there cannot be more than one value s_i that the sender can reveal without finding a collision under the hash function. Hence, if two distinct values for s_i can generate the same Merkle root, then either the hash function is not properly collision-resistant, or one of the computations went wrong.

9.4 MERKLE TREE – ATTRIBUTE-BASED ENCRYPTION

Merkle Tree – Attribute-Based Encryption (MT-ABE) is an algorithm that combines the advantages of both ABE and Merkle trees. It considers the RBAC capabilities of ABE, along with the verifiability and string commitment properties of the Merkle tree to generate a post-quantum resistant access control mechanism.

MT-ABE makes use of a pre-computed Merkle subtree along with a random token to ensure authentication and authorization. It works in similar lines to CP-ABE, which means that the key is common for everyone, but decrypting the information is regulated by a set of user attributes. In addition, a random token has to be generated and communicated between server and client, which may be implemented by a suitable authentication app like Microsoft Authenticator, or via simple One Time Password.

9.4.1 DESCRIPTION OF THE ALGORITHM

The working of MT-ABE can be split into two phases. The first phase can be mentioned as a **setup** phase, where appropriate rules are set, and doctors get themselves registered on the platform. The second phase is the **access** phase, where access request takes place. It can be noted that the setup phase is required only once, while the access phase is repeated every time. The setup phase proceeds as below.

1. Patient's record is already available on the EHR Server. Patient can set specific access rules like (DESIGNATION=CARDIOLOGIST) AND (HOSPITAL=CITY HOSPITAL). The rules are set via an online form and dispatched to the server.
2. A Merkle subtree is generated based on the rules. The root of the subtree is stored in the server for future verification.
3. Doctor has to register with the server using his information. Based on his information, a Merkle tree is generated, and the root is stored as an ID subtree in the server.

The access phase is illustrated in Figure 9.3.

The phase proceeds as follows:

1. The doctor logs in to the portal and issues a new request.
2. The server generates an authentication token and sends it to the doctor using an OTP or through an authenticator app. In addition, the server requests for information from the portal based on the requirements laid out in the access rule.
3. Based on the requirements and token, a Merkle tree is generated, which should match the one in the server in order to gain access (that is verified in the coming steps). The ID subtree is combined with the random token to complete the ID Merkle tree. The Merkle root and ID root are sent to the server for further verification.
4. The server regenerates the Merkle tree using the Merkle subtree already available with it and the token that was sent to the doctor.
5. If both trees match, it means that the requirements and token are correct, and access can be granted. In that case, the data is decrypted using its key that is stored in the server, and a response is issued. If the Merkle roots don't match, an error response is returned.

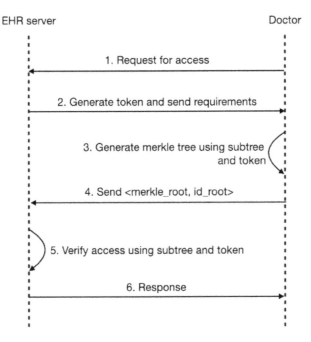

EHR server   Doctor

1. Request for access

2. Generate token and send requirements

3. Generate merkle tree using subtree
   and token

4. Send <merkle_root, id_root>

5. Verify access using subtree and token

6. Response

FIGURE 9.3 Access phase for MT-ABE

9.4.2 GENERATING MERKLE SUBTREE USING ATTRIBUTES

Assume an EHR is defined by the patient to be restricted to only physicians in St. Johns Hospital, Atlanta. Its rule will be written as:

(DESIGNATION = PHYSICIAN) AND (HOSPITAL = ST.JOHNS) AND
   (LOCATION = ATLANTA)

It can be converted to JSON form as follows. Each strict requirement is coded into a separate JSON entry.

{DESIGNATION: PHYSICIAN}
{HOSPITAL: ST.JOHNS}
{LOCATION: ATLANTA}

These form inputs to the Merkle subtree, which is generated as per Figure 9.1.

D1 = {DESIGNATION: PHYSICIAN}
D2 = {HOSPITAL: ST.JOHNS}
D3 = {LOCATION: ATLANTA}
H1 = SHA256(D1)
H1 = 9779aca36dc9ec44be4a33df5369e5088b091bb228a5fd7777407dc-
   17259ceee

$H2 = SHA256(D2)$

$H2$ = f3e2ec0dc9680be03b2aa351ef0e43c5f942aa67bf0eb833a57d83d-91ba56a38

$H3 = SHA256(H1H2)$

$H3$ = 7b62924e620b33a1a82f37e0628b855d4e6cf534cc44ee59fc6d-774b362a88a0

$H4 = SHA256(D3)$

$H4$ = d0c69f3b1d7a6f1a1ef540e16377df3a34df512f531badd01430afe8d6ec217c

$H5 = SHA256(H3H4)$

$H5$ = c8c798d16e954378da58e2b08bd8fd42d6b87bc0aca0b95cc2079a-b00e80eaf8

This will be the Merkle subtree for the required access control. The root will contain the value of H5. It can be observed that the result will be a skewed Merkle tree, designed to limit the number of children to two for any node. Due to the inherent nature of the Merkle tree, only the root of the subtree would be required to be stored in the server as part of access control. The tree can be visualized as shown in Figure 9.4.

9.4.3 AUTHENTICATION AND AUTHORIZATION USING MT-ABE

Authentication takes place with the help of the token that is generated by the server. Since we use cryptographic hash functions like SHA-256, any minor change in the input will cause a huge change in the output. When the server issues the token, the portal at the doctor's end will generate two trees.

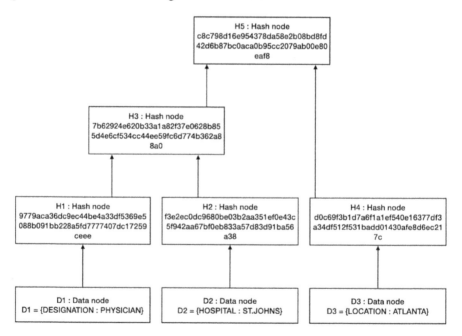

FIGURE 9.4 Merkle subtree for (DESIGNATION = PHYSICIAN) AND (HOSPITAL = ST. JOHNS) AND (LOCATION = ATLANTA)

1. ID Merkle subtree, which stores the entire profile information of the doctor for verification purposes. This subtree is generated and stored in the server during registration itself, and during the authentication phase, this subtree is generated from scratch in the client and combined with the token to form the final ID Merkle tree root.
2. The access control Merkle tree, which is based on the requirements laid out by the patient. The tree is generated at the doctor's portal by filtering out only the necessary information requested by the server, in the same order as requested, and finally combined with the token to form the Merkle tree root.

For authentication, the ID Merkle tree root is taken in the server. The server already contains the ID subtree and token sent across to the doctor. The tree is completed using these two pieces of information and the root is verified with the root received from the doctor. If they match, it is authenticated.

For authorization, the access control Merkle tree generated by the doctor is taken. The EHR is already tagged along with its access control subtree. This subtree is taken and combined with the token to generate the root. This root is verified with the one obtained from the doctor to ensure authentication.

The advantage of this method is manyfold.

1. The key is always kept in the server. Only RBAC is used to keep unauthorized access away.
2. The doctor's information never goes to the server during access request. It is sent only during registration. All further steps involve only verification.
3. The patient's access control rules never leave the server. The server only broadcasts the required fields, and the values are later verified.
4. The token is able to mask both authentication and authorization data using a quantum-safe mechanism, removing possibilities of any reverse engineering.

9.5   DEMONSTRATION OF THE ALGORITHM

Let us consider an example to see how the end-to-end process works in MT-ABE.

9.5.1   Setup Phase

AN EHR is tagged with patient P1, and P1 decides that only cardiologists in Lenox Hill Hospital, New York can access the data. The rule will be (DESIGNATION=CARDIOLOGIST) AND (HOSPITAL=LENOX HILL) AND (LOCATION=NEW YORK). Based on this, an access control Merkle subtree is generated as follows.

D1 = {DESIGNATION: CARDIOLOGIST}
D2 = {HOSPITAL: LENOX HILL}
D3 = {LOCATION: NEW YORK}
$H1 = SHA256(D1)$
H1 = 3e8ceee8ce5e2c06969762aad77f31089dab97262477e6e0246a24eb
45765188

$H2 = SHA256(D2)$

H2 = 94fb05cb76a4f7ab1dafd2cd991a0f71991d008c49d55a753d8e15c144115 d2a

$H3 = SHA256(H1H2)$

H3 = 45b46f42a6eca5957775fb87e8281b4624f48bb27faee39c049d71c754a62 ba0

$H4 = SHA256(D3)$

H4 = 88947252f0e000a8553f540e01fbc7b138537a3fce88c58ffdb1566b4ff5 ea08

$H5 = SHA256(H3H4)$

H5 = 28b7574616b0686272daa8e227b3e237e3fdbe5604e0ceac4a 22c543dd3194e4

Thus, the server will contain one entry in its database, which is

{Patient_ID: P1,
EHR_ID: 1,
ATTRIBUTE_LIST: [DESIGNATION, HOSPITAL, LOCATION],
ENCRYPT_KEY: e80c451d5e9c1df8c4b1b2f3f1943a2a,
ROOT:
28b7574616b0686272daa8e227b3e237e3fdbe5604e0ceac4a22c543dd
3194e4}

In order to access a record, a doctor should be registered on the portal. Based on all his records, a tree is generated in a similar fashion, and the root is placed on the server's database. A typical entry would look like the following.

{DOC_REGN_NO: 14465,
NAME: SAM THOMAS,
LOCATION: NEW YORK,
HOSPITAL: LENOX HILL,
DESIGNATION: CARDIOLOGIST,
ROOT:
3381065d936564daf8d4e9119dddfc3116657b9553c89d9d6284dbeca91cf370,
LATEST_TOKEN: 0}

This process is illustrated in Figure 9.5.

9.5.2 Access Phase

When doctor 14465 wants to access EHR 1, a request is generated for the same. The server responds by generating a token and sending it across to the doctor, with all additional requirements for authorizing the access.

Assume the token 4949536 is generated. This is reflected in the database as follows.

{DOC_REGN_NO: 14465,
NAME: SAM THOMAS,

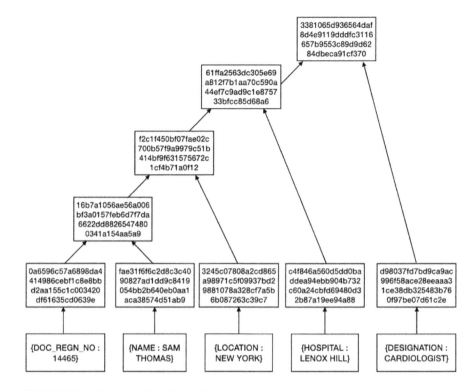

FIGURE 9.5 Example of an ID Merkle tree

LOCATION: NEW YORK,
HOSPITAL: LENOX HILL,
DESIGNATION: CARDIOLOGIST,
ROOT:
3381065d936564daf8d4e9119dddfc3116657b9553c89d9d6284dbe-
  ca91cf370,
LATEST_TOKEN: 4949536}

This is sent via OTP or via an authentication app. Then, the server looks up the list of requirements from the EHR database and finds [DESIGNATION, HOSPITAL, LOCATION]. Thus, it sends the response:

REQUIRED: [DESIGNATION, HOSPITAL, LOCATION]

Once the doctor's portal receives the token and these requirements, the following are done.

1. The ID Merkle subtree is regenerated from scratch.
2. The requirements are looked up, and subtree is generated.
3. Once the token is received, it is entered by the doctor in their portal, which is hashed and combined with both subtrees to generate the Merkel roots.

In this example, when Step 1 is completed, the same subtree as in Figure 9.5 is created, and the root value will be

3381065d936564daf8d4e9119dddfc3116657b9553c89d9d6284dbeca91cf370. Then, the token is combined with this subtree as follows.

H1 = 3381065d936564daf8d4e9119dddfc3116657b9553c89d9d6284dbe-
    ca91cf370
TOKEN = 4949536
H2 = SHA256(TOKEN)
H2 = 03487536f12214060fcbc2824806b0f46452dbb09aa814948ff23888d-
    5bbd185
root = SHA256(H1H2)
root = f1c0a1dbd9fdf4bffaa81eabd83e3a9652cd69fa8d4bdd1ae695947a3e0
    b90f8

Similarly, the requirements are looked up, and the Merkel subtree is generated. For that, the JSON data is built from the list of requirements, and their values available in the portal. So we get: REQUIRED: [DESIGNATION, HOSPITAL, LOCATION] {DESIGNATION: CARDIOLOGIST, HOSPITAL : LENOX HILL, LOCATION: NEW YORK}

This is split into three data entries and the Merkle tree is built as follows.

D1 = {DESIGNATION: CARDIOLOGIST}
D2 = {HOSPITAL: LENOX HILL}
D3 = {LOCATION: NEW YORK}
H1 = SHA256(D1)
H1 = 3e8ceee8ce5e2c06969762aad77f31089dab97262477e6e0246a24eb
    45765188
H2 = SHA256(D2)
H2 = 94fb05cb76a4f7ab1dafd2cd991a0f71991d008c49d55a753d8e15c144115
    d2a
H3 = SHA256(H1H2)
H3 = 45b46f42a6eca5957775fb87e8281b4624f48bb27faee39c049d71c754a62
    ba0
H4 = SHA256(D3)
H4 = 88947252f0e000a8553f540e01fbc7b138537a3fce88c58ffdb1566b4ff5
    ea08
H5 = SHA256(H3H4)
H5 = 28b7574616b0686272daa8e227b3e237e3fdbe5604e0ceac4a22c543
    dd3194e4

This is then combined with the token to get the following.

H1 = 28b7574616b0686272daa8e227b3e237e3fdbe5604e0ceac4a22c
    543dd3194e4
TOKEN = 4949536

H2 = SHA256(TOKEN)
H2 = 03487536f12214060fcbc2824806b0f46452dbb09aa814948ff23888d-
5bbd185
root = SHA256(H1H2)
root = bb830e2a2bbd37714c40109b1aa07f6d4a6376fc5bad-
b6493473bd0affbf8d76

Once both trees are completed, the roots are sent to the server.

{
MERKLE_ROOT:
bb830e2a2bbd37714c40109b1aa07f6d4a6376fc5bad-
b6493473bd0affbf8d76,
ID_ROOT:
f1c0a1dbd9fdf4bffaa81eabd83e3a9652cd69fa8d4bdd1ae695947a3e0b90f8
}

The server now starts the verification. It first authenticates the doctor. For that, it takes the ROOT value in doctor's database and merges it with the LATEST_TOKEN value.

ROOT = 3381065d936564daf8d4e9119dddfc3116657b9553c89d9d6284dbe-
ca91cf370
LATEST_TOKEN = 4949536
H = SHA256(LATEST_TOKEN)
H = 03487536f12214060fcbc2824806b0f46452dbb09aa814948ff23888d-
5bbd185
VERIFY = SHA256(ROOT.H)
VERIFY =
f1c0a1dbd9fdf4bffaa81eabd83e3a9652cd69fa8d4bdd1ae695947a3e0b90f8
ID_ROOT =
f1c0a1dbd9fdf4bffaa81eabd83e3a9652cd69fa8d4bdd1ae695947a3e0b90f8

Since they are identical, the doctor is authentic. In this step, the server was able to verify that the entire information being made available in the portal is exactly as per the information available in the server (doctor's portal generated the ID tree from scratch, while the server had it pre-computed).

Next, the authorization is checked. For that, the ROOT value in EHR database is taken and merged with the LATEST_TOKEN value.

ROOT=28b7574616b0686272daa8e227b3e237e3fdbe5604e0ceac4a22c543dd
3194e4
LATEST_TOKEN = 4949536
H =SHA256(LATEST_TOKEN)
H=03487536f12214060fcbc2824806b0f46452dbb09aa814948ff23888d5bbd185
VERIFY = SHA256(ROOT.H)

VERIFY =

b b 8 3 0 e 2 a 2 b b d 3 7 7 1 4 c 4 0 1 0 9 b 1 a a 0 7 f 6 d 4 a 6 3 7 6 f c 5 b a d -
b6493473bd0affbf8d76

MERKLE_ROOT=

b b 8 3 0 e 2 a 2 b b d 3 7 7 1 4 c 4 0 1 0 9 b 1 a a 0 7 f 6 d 4 a 6 3 7 6 f c 5 b a d -
b6493473bd0affbf8d76

Since they are identical, the doctor is now authorized to access. This is possible because the Merkle tree from the doctor was generated based on the requirement list and values available in the portal (which was just verified to be correct as per records), and the root value in the server was pre-computed from patient requirements. Once the authorization is verified, the value of ENCRYPT_KEY is fetched from the EHR database and the data is decrypted for transferring to the portal.

If we consider a counter-example where the doctor is a pulmonologist, the authentication will still be successful, but the MERKEL_ROOT will be generated as follows.

D1 = {DESIGNATION: PULMONOLOGIST}

D2 = {HOSPITAL: LENOX HILL}

D3 = {LOCATION: NEW YORK}

H1 = SHA256(D1)

H1=b408bd2ab555bdb552a84ab063fc1730743721bc7558c8eafa519bf73d
b1b890

H2 = SHA256(D2)

H2=94fb05cb76a4f7ab1dafd2cd991a0f71991d008c49d55a753d8e15c144115
d2a

H3 = SHA256(H1H2)

H3=f49f34812876eecdd2e5a9dff93a7fb2412c3a129361c922a74e62e9
06e43874

H4 = SHA256(D3)

H4 = 88947252f0e000a8553f540e01fbc7b138537a3fce88c58ffd-
b1566b4ff5ea08

MERKLE_ROOT = SHA256(H3H4)

MERKLE_ROOT =

8158831a7471f19a94eb64403599c93f1ac2fdc13f9250c9b838fd9e656f400a

ROOT=28b7574616b0686272daa8e227b3e237e3fdbe5604e0ceac4a22c543dd
3194e4

LATEST_TOKEN = 4949536

H = SHA256(LATEST_TOKEN)

H = 03487536f12214060fcbc2824806b0f46452dbb09aa814948ff23888d-
5bbd185

VERIFY = SHA256(MERKLE_ROOT.H)

VERIFY =

b b 8 3 0 e 2 a 2 b b d 3 7 7 1 4 c 4 0 1 0 9 b 1 a a 0 7 f 6 d 4 a 6 3 7 6 f c 5 b a d -
b6493473bd0affbf8d76

MERKLE_ROOT =

8158831a7471f19a94eb64403599c93f1ac2fdc13f9250c9b838fd9e656f400a

It can be seen that they don't match when atleast one entry is different from the required list.

9.6 PERFORMANCE MEASURES

The performance of this method can be measured on two levels – quantum safety and collusion resistance.

MT-ABE is inherently quantum-safe [8,20]. This is since cryptographic hash functions are nearly impossible to reverse engineer. Even with a quantum computer, the best chance of breaking the hash is a brute force search, which will take 2^{256} iterations for SHA-256. If we use hash functions with larger hash sizes, the chances are even less. In addition, the Merkle tree is built not on top of just one hash, but on multiple levels of hashes. The architecture for MT-ABE has been built in such a way that it is a skewed tree, requiring close to N intermediate hashes itself for N values. This exponentially increases the difficulty.

MT-ABE doesn't disclose much information during access. The information sent across the channel is:

1. Access request
2. List of requirements from the server
3. ID Merkle tree and access Merkle tree
4. Final response

Even if an attacker gains access to the list of requirements or both Merkle trees, there is no way they can be broken to reveal the original access details. In addition, even if the token is intercepted, there won't be many benefits because the string commitment property of Merkle trees ensures that one part of the data can be safely exposed without affecting any other.

Finally, MT-ABE is collusion resistant. Even if multiple doctors collude to escalate their access privileges, the authentication step blocks any attempt to use a different set of privileges other than the ones already given to the doctor [5].

9.7 SUMMARY AND CLOSING REMARKS

The chapter was aimed at presenting an authentication and authorization model based on Merkle trees and attribute-based encryption. Merkle trees are quantum-safe, while attribute-based encryption enforces proper role-based access control. Together, the advantages are combined to form Merkle Tree – Attribute-Based Encryption (MT-ABE), which was demonstrated to work well under practical circumstances. It has also been mentioned that the hybrid algorithm keeps all properties of its original algorithms intact. Hence, this is a good candidate to be deployed in the domain of Healthcare 4.0, where automation and personalization of health is the norm.

REFERENCES

1. Attrapadung, N., Libert, B., & Panafieu, E. D. (2011). Expressive key-policy attribute-based encryption with constant-size ciphertexts. In: *International Workshop on Public Key Cryptography*, pp. 90–108.

2. Barrows Jr, R. C., & Clayton, P. D. (1996). Privacy, confidentiality, and electronic medical records. *Journal of the American Medical Informatics Association*, 3 (2), 139–148.
3. Bayardo, R. J., & Sorensen, J. (2005). Merkle tree authentication of http responses. In: *Special Interest Tracks and Posters of the 14th International Conference on World Wide Web*, pp. 1182–1183.
4. Benaloh, J., Chase, M., Horvitz, E., & Lauter, K. (2009). Patient controlled encryption: Ensuring privacy of electronic medical records. In: *Proceedings of the 2009 ACM workshop on Cloud computing security*, pp. 103–114.
5. Bethencourt, J., Sahai, A., & Waters, B. (2007). Ciphertext-policy attributebased encryption. In: *2007 IEEE symposium on security and privacy (SP'07)*, pp. 321–334.
6. Chalkias, K., Brown, J., Hearn, M., Lillehagen, T., Nitto, I., & Schroeter, T. (2018). Blockchained post-quantum signatures. In: *2018 IEEE International Conference on Internet of Things (iThings) and IEEE Green Computing and Communications (GreenCom) and IEEE Cyber, Physical and Social Computing (CPSCom) and IEEE Smart Data (Smart- Data)*, pp. 1196–1203.
7. Chelladurai, U., & Pandian, S. (2021). Hare: A new hash-based authenticated reliable and efficient modified Merkle tree data structure to ensure integrity of data in the healthcare systems. *Journal of Ambient Intelligence and Humanized Computing*, 12 (4), 1–15.
8. Chen, L., & Movassagh, R. (2021). Quantum Merkle trees. *arXiv preprint arXiv:2112. 14317.*
9. Ederov, B. (2007). *Merkle tree traversal techniques.* Bachelor Thesis, Darmstadt University of Technology Department of Computer Science Cryptography and Computer Algebra.
10. El Gafif, H., & Toumanari, A. (2021). Efficient ciphertext-policy attribute-based encryption constructions with outsourced encryption and decryption. *Security and Communication Networks*, 2021, 8834616.
11. Hülsing, A., Gazdag, S.-L., Butin, D., & Buchmann, J. (2015). Hash-based signatures: An outline for a new standard. In: *NIST Workshop on Cybersecurity in a Post-Quantum World*, pp. 1–12.
12. Hoerbst, A., & Ammenwerth, E. (2010). Electronic health records. *Methods of Information in Medicine*, 49 (4), 320–336.
13. Kuhn, D. R., Coyne, E. J., Weil, T. R., et al. (2010). Adding attributes to role-based access control. *Computer*, 43 (6), 79–81.
14. Li, H., Lu, R., Zhou, L., Yang, B., & Shen, X. (2013). An efficient Merkle-treebased authentication scheme for smart grid. *IEEE Systems Journal*, 8 (2), 655–663.
15. Li, J., & Carayon, P. (2021). Health care 4.0: A vision for smart and connected health care. *IISE Transactions on Healthcare Systems Engineering*, 11 (3), 171–180.
16. Liu, J., Li, X., Ye, L., Zhang, H., Du, X., & Guizani, M. (2018). Bpds: A blockchain based privacy-preserving data sharing for electronic medical records. In: *2018 IEEE Global Communications Conference (GLOBECOM)*, pp. 1–6.
17. Meng, F., Cheng, L., & Wang, M. (2021). Ciphertext-policy attribute-based encryption with hidden sensitive policy from keyword search techniques in smart city. *EURASIP Journal on Wireless Communications and Networking*, 2021 (1), 1–22.
18. Miller, A. R., & Tucker, C. (2009). Privacy protection and technology diffusion: The case of electronic medical records. *Management Science*, 55 (7), 1077–1093.
19. Mohan, A. P., & Gladston, A. (2020). Merkle tree and blockchain-based cloud data auditing. *International Journal of Cloud Applications and Computing (IJCAC)*, 10 (3), 54–66.
20. Naor, M., & Yung, M. (1989). Universal one-way hash functions and their cryptographic applications. In: *Proceedings of the twenty-first annual ACM symposium on Theory of computing*, pp. 33–43.

21. Nguyen, D.-M., Luu, Q.-H., Huynh-Tuong, N., & Pham, H.-A. (2019). MB-PBA: Leveraging Merkle tree and blockchain to enhance user profile-based authentication in e-learning systems. In: *2019 19th International Symposium on Communications and Information Technologies (ISCIT)*, pp. 392–397.
22. Sandhu, R. S. (1998). Role-based access control. In: *Advances in Computers, 46,* pp. 237–286.
23. Sharma, B., Sekharan, C. N., & Zuo, F. (2018). Merkle-tree based approach for ensuring integrity of electronic medical records. In: *2018 9th IEEE Annual Ubiquitous Computing, Electronics & Mobile Communication Conference (UEMCON)*, pp. 983–987.
24. Tortorella, G. L., Fogliatto, F. S., Mac Cawley Vergara, A., Vassolo, R., & Sawhney, R. (2020). Healthcare 4.0: Trends, challenges and research directions. *Production Planning & Control*, 31 (15), 1245–1260.
25. Wang, C., & Luo, J. (2013). An efficient key-policy attribute-based encryption scheme with constant ciphertext length. *Mathematical Problems in Engineering*, 2013, 810969.
26. Waters, B. (2011). Ciphertext-policy attribute-based encryption: An expressive, efficient, and provably secure realization. In: *International workshop on public key cryptography*, pp. 53–70.
27. Yin, H., Xiong, Y., Zhang, J., Ou, L., Liao, S., & Qin, Z. (2019). A keypolicy searchable attribute-based encryption scheme for efficient keyword search and fine-grained access control over encrypted data. *Electronics*, 8 (3), 265.

10 Artificial Intelligence, Machine Learning and Smartphone-Internet of Things (S-IoT) for Advanced Student Network and Learning

Santosh R. Gaikwad
MET's Institute of Management

CONTENTS

10.1 INTRODUCTION

Artificial Intelligence (AI) can be described as the ability of a computer system or robot which is controlled by a computer to perform any kind of activity that is usually done by humans because it requires intelligence and judgment capacity (Chen et al., 2022). It can be said that artificial intelligence helps in smart decision making and provides ample opportunities to the user for research and data analysis. In general terms, AI also manages repetitive tasks that can cause boredom to humans. From a student's perspective, it can be said that AI is useful to them to answer the most commonly asked questions in a very short time with the help of automation and conversational intelligence.

- In the coming years, the concept of smart classrooms will be more popular among students for vocational courses and colleges.
- A smart classroom is a technology-based learning approach that can help the students to improve their skills even without visiting a traditional classroom.

DOI: 10.1201/9781003250357-10

- The popularity of smart classrooms has increased after the pandemic because at that time it became mandatory for the students to attend classrooms from home.

It also enables the educators to create interactive experiences at the time of teaching and for this reason, the smart classroom, which AI-based, enable the students to become more appealing among students.

AI can be used for advanced student networks and learning to provide an instructional environment for the students and teachers to focus on the usage of technology (Hwang and Tu, 2021). With the help of AI, the teachers can provide smart teaching and it can improve conventional learning outcomes. An appropriate implementation of AI in the classroom helps in providing customized solutions to the students to focus on research abilities and it also boosts the learning efficiency of the students. This technology is useful to the students because even without a teacher the students can solve their queries and analyze the outcomes based on the data volume. The main objective of using AI in the classroom is to make the students more successful in curriculum outcomes and the three most popular learning techniques such as mnemonic, structural and generative can be used by the students with the help of AI (Mishra et al., 2020).

It is important for the teachers to use AI to formulate strategies that are practically crucial in the case of Marvel education because a huge amount of resources for active automation interaction is required to communicate with the student. AI also helps the students with learning strategies to improve their concentration and it ultimately makes the student more conscious of their learning patterns (Zawacki-Richter et al., 2019). AI also helps the students to become more independent about their learning strategies.

In today's learning environment, the popularity of vocational education is continuously increasing because it provides great opportunities for learners to improve their skills and career. Technical and vocational education training is important for economic progress and improvement in the efficiency of the employees. AI is an indispensable factor in providing vocational training in today's world. E-learning technology is helping the students to continue their education along with their careers. The teachers can also take the help of AI by using the *Artificial Neural Network* for the prediction of the final performance of the student in the case of the online learning platform (Mishra et al., 2020). It is an important method to forecast the students' performance in the e-learning environment used in the learning management system.

It is found that the researchers from Stanford in association with Google and NVIDIA succeeded in creating the largest artificial neural network. It is six times bigger in size than any other normal unit built. It is created exclusively for machine learning purposes. Figure 10.1 shows how artificial neural network can be applied in several areas such as web search, recognition of objects at ease, speech recognition, translation from one to many languages and self-driving cars. With respect to the field of education, speech recognition and language translation have contributed to a great extent in minimizing the existing barriers and challenges for students. For instance, non-native speakers can use speech and language translation tools to learn the native variety of English. It plays a significant role in the teaching and learning

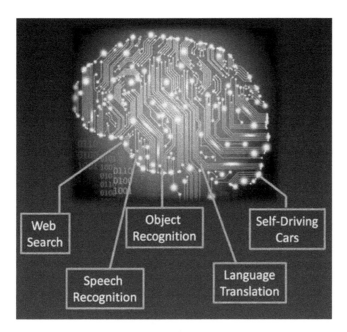

FIGURE 10.1 Use of artificial neural network (Source: Scoop.It.)

of English as a Second Language (ESL) and English as a Foreign Language (EFL). The scope can be extended to content management systems for product descriptions, brochures, marketing material, business policies, creating content in target language and translating it into several local languages at the same time. It leads to providing customized solutions for the products and services aiming for high-level stakeholder satisfaction.

Donald Clark (2016) has a great deal of contribution to the field of AI. His 30 years of profound experience paved the way to explore new opportunities in the world of education, transforming teaching and learning processes. While addressing as a keynote speaker at an international conference, he explained how bad learners can gain the competitive advantages offered by AI. The inattentiveness, demotivation, sleep, lack of networking opportunities, anxiety, cognitive overload, tendency to forget and social as well as cognitive biases restrict the scope for bad learners. He commented on the limitations of the human brain stating that the brain dies after several years, with no provision for downloading the ideas and memory. However, AI preserves the knowledge for hundreds of years. The fascinating ideas and network intelligence make it possible to think and act beyond human brain capacity. Furthermore, it enables the bad learners to overcome these challenges and develop interest slowly and steadily through advanced features such as gamification, visualization and others. It is useful to create interactive, gamified and simulation-based courses for the teachers as well. Though it requires expertise, it leads to automation saving time, money, labor and increasing the efficiency of the teaching-learning program altogether.

Another model which is used in the case of e-learning environment is Tracking Student learning and Knowledge (TSLAK) model. This model has two variables monitoring the student learning and the student knowledge. With the use of this model, the teacher can also understand the knowledge acquisition process of the students and understand the reason for the significant decrease in the interest of the students. The AAI learning model can be used in an interactive education system mainly for vocational courses. In today's business environment, industries phase demand for skilled workers and the pressure of increasing cost for providing training to the employees is continuously increasing. In this scenario, AI is a blessing for skilled labor to enhance their knowledge and become competent as per the industry's demand (Guan et al., 2020). The learning process in the e-learning platform is related to the goals and skills of the students and the learning process can take place on an informal basis and in a less unplanned way. With the help of AI, the student can discover new skills and knowledge independently.

The main benefit of using AI learning platforms is personalization, which is becoming one of the biggest trends in education. This technology is beneficial to the students to adopt a personalized approach to the learning programs and the students can also use their preferences and unique experiences while learning something new. AI can adapt according to the knowledge level of the student and learning speed (Malik et al., 2019). Apart from that, AI can also analyze the previous history of learning of the students and identify the weaknesses and offer the best courses that can help the students to improve their knowledge and provide many opportunities. Different chatbots powered by AI can provide extra guidance to the students as it is a perfect solution for the students to sharpen their skills and improve weaknesses outside the classroom (Haristiani, 2019). It can also be said that this modern technology provides one-on-one learning experience to the students and it is going to be a perfect replacement for an educator. Another major benefit of using AI solutions in e-learning platforms is that it provides quick responses to the questions asked by the students. The students can find their required solution in the most asked questions section with the help of conversational intelligence and the support of automation. It helps the students to track down the answers quickly without waiting a long time for the answer. The students can also access online courses anytime from anywhere and each student can learn at their own pace.

This work is organized in the below mentioned three sections:

- The first section presents the introduction and broader view of various latest technologies used for the student network and to optimize the effectiveness of learning.
- The second section illustrates how machine learning can be used effectively for the advanced student network and learning.
- The last section demonstrates the scope of applying the Smartphone-Internet of Things (S-IoT) as a platform for student network and learning.

10.2 MACHINE LEARNING FOR ADVANCED STUDENT NETWORK AND LEARNING

Machine Learning (ML) can be described as a method of analyzing the data automatically and it can be regarded as a part of AI that is based on an analytical model and identifies the patterns from the data (Shah et al., 2021). The main benefit of using

ML is that even after exposure to the new data, the computer can independently adopt that by learning from the previous computation and repeatable decisions. Even though ML algorithms were present around for a long time, the format of complex mathematical calculations for big data is a recent innovation. So, it can be said that ML can bring a revolutionary change to the educational sector if it is properly applied.

ML can be used to transform education and change the teaching-learning and research pattern fundamentally. The teachers can use ML to support the weakest student as early as possible and modify their actions to improve the success rate. In the case of higher studies and research, ML is also important for unlocking new insights and discoveries (Mishra et al., 2020). Apart from that, ML can also be used to expand the research process and impact the online learning environment through transcription, personalization, localization and text-to-speech. In the case of Higher Education, ML can be used to identify the right student, predict the outcome of the student performance, enrollment of the students and ensure the success of the students (Giannakos et al., 2020). However, it is necessary to accelerate the research on ML to make it less costly so that the application of it can be done on a broader prospect. ML can also be used to modernize the experience on campus by making them smarter, more efficient and safer.

ML can not only be used in higher education but also it can be efficiently implemented in primary schools. It can be used to identify at-risk students and enhance campus security with the ML-powered threat detection process. In primary school, ML can be used to improve the teacher's efficiency by making personalized content for the students. In the case of remote locations, ML can be used as a teaching assistant and replacement of the education to teach the students and improve the assessment process and grading system. As life is getting busy for the parents, ML can be used to provide self-service tools for both students and parents (Iatrellis et al., 2020). ML can also be used to reach global companies because with the help of this technology, the learners can take benefit from personalized learning content, educational chatbots and self-service capabilities. Different learning companies can use this technology for targeting the students and tracking and predicting the performance of the students (Luckin, 2018). It can be said that ML in the educational sector can be used for a personalized experience to provide each student with an individualized educational platform and help them in their own learning.

ML is bringing a revolution in the education sector by introducing adaptive learning, increasing the efficiency of the students as well as teachers, Learning Analytics, Predictive Analytics, personalizing content and accurately grading the assignment. Adaptive learning can be described as a self-explanatory method of learning. This technology can be used to analyze the performance of the students on a real-time basis and make modifications in the teaching pattern to adapt to the students' learning capabilities. This advanced technology also helps the student to choose the learning path which will be most suitable for them. The students also get suggestions from various learning materials and make the best decisions for them. It can be said that ML has the capability to offer the best quality curriculum management process and it is also useful to bifurcate work according to the potential of the students (Iatrellis et al., 2020). It makes the educator and students more comfortable in the learning environment so that they can work collaboratively to get the best outcome. ML is also necessary to increase the engagement of the students in learning to improve their educational performance. This software is also useful in scheduling activities and classroom management.

With the help of ML technology, teachers can dive deep into data and simplify it for the students. The teachers can analyze various contents and make appropriate conclusions from them. Learning analytics also helps the student by suggesting them the most appropriate path and learning methodologies for them (Iatrellis et al., 2020). Another most important feature of ML is predictive analysis. It helps the teachers to understand the students' mindset and specific needs of the students. It also helps the faculty of a learning institution and parents to make appropriate decisions to get the best outcome for the students. ML can also be used for assessing the students' performance more accurately than a human can perform. This technology provides valid and reliable analysis of the student's performance by lowering the chances of error. A personalized learning approach is useful to make a tailor-made pathway for the individual students and spot their issues (Conati et al., 2018).

The educational institutions which are using ML can provide a great campus experience to the students by providing them self-service capabilities. It can be observed that traditionally group discussions, video lectures and different kinds of platforms of online learning content were not so cost-effective due to translation and transcription at scale. However, these days it can be observed that due to deep learning power translation, the students can easily reach the online learning content because it has become a low-cost option for them (Luckin, 2018). ML is also useful for the large educational institution in case of admission of the students and forecasting the enrollment process for optimizing the capacity of the institution. ML can also be used in data protection and to prevent fraud (Temitayo Sanusi, 2021). However, there are several disadvantages of using ML in education such as data acquisition and time and resources. In order to appropriately use ML, it is required to acquire a huge data set and this must be unbiased and of good quality. So, for small learning institutions, it is not possible to take the advantage of ML on a full scale.

The recent developments in ML have transformed the business world, learning places and communities by and large. It is possible to monitor the human behavioral patterns and direct them towards the guided exercises and activities for better outcomes which are predefined in the forms of learning objectives for any course. It further enables the course administrators for formative assessments, peer collaboration of students, cognitive learning and human-to-human or human-to-computer level interactions as well. The feedback mechanism is another strength experienced by the learners during the course.

ML proves to be a networking tool to integrate the various functions and tasks related to the instructors, learners, assessors and administrators of the course. This is critical in the case of a large number of students and formative assessments. As shown in Figure 10.2, ML serves as a tool or platform wherein input, process and outcomes are well integrated. It enlarges the scope of accountability, transparency, automation and database management to the great extent. It syncs the various sub-verticals and provides integrated solutions. The instructor can get a better idea about the learning styles to improvise the performance in coming sessions.

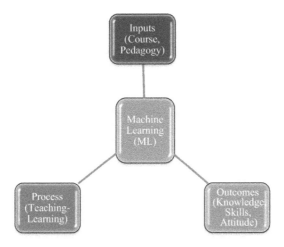

FIGURE 10.2   Machine learning as a tool for teaching-learning

10.3   SMARTPHONE-INTERNET OF THINGS (S-IoT) FOR ADVANCED STUDENT NETWORK AND LEARNING

IoT is described as a network of various physical objects that are connected with different technologies, software and sensors for exchanging data through the internet (Ramlowat and Pattanayak, 2019). IoT devices refer to the hardware that is programmed to transmit data over the internet. In the educational sector, IoT in schools provides a better connection and collaborative learning environment for the students. This technology also offers better access to the learning materials and various communication channels to the students. Teachers can also measure the learning progress of the students on a real-time basis. IoT devices such as smartphones are changing the way a student can observe and process data for better outcomes (Luckin, 2018).

Most parents are very much optimistic about the smart classroom and its impact on their child's life, and they believe that IoT can improve the learning environment for a better cause. As the students are more acquainted with the usage of smartphones, it will be easy to use IoT technologies in school and different types of learning institutions. IoT technologies are beneficial in the educational sector to automate attendance, properly locate and record the student's location on the campus, block the exit if a student wants to live at an odd time and create digital ID cards. Apart from that, the administration staff in the educational institutions can also use this technology to control the infrastructure such as lighting and maintenance of school equipment from their smartphones.

IoT also facilitates a cashless environment in the school. The students can pay the school fees, without carrying cash, from their smartphones (Chweya et al., 2020). The popularity of this technology has increased after the pandemic because online teaching has become an indispensable factor now. It becomes quite a challenging factor for the parents to separate their kids from smart devices such as smartphones, and it can be observed that their screen time has increased significantly. As the children are getting more acquainted with smartphones, it has become a usual choice to provide education with the help of this technology. IoT devices use different sensors and software in addition to smartphones for the creation of a smart classroom.

The IoT helps the students to use their smartphones and login into the learning software of the schools from anywhere. Smartphones can be the best virtual notebook and test book for the students if this technology is properly used. In order to improve their learning skills, the students can mark the important paragraph, highlight the text and make little notes on their textbook (Luckin, 2018). One of the major benefits of this technology is that students can learn in many ways beyond the traditional way by moving towards smart devices and learning through debates, webinars, online discussions and group activities which are all administered through the internet. One of the main examples of IoT technology is Promethean. It is designed for interactive display of natural writing technology and dry erase features. Promethean also provides cloud-based lessons to the students to enhance their knowledge. Different schools have integrated IoT technology for tutoring various subjects and media centers to communicate with the students and provide them with the required study materials (Maksimović, 2018).

After the popularity of various social media platforms, it has been observed that the attention of the students has significantly decreased in recent times. IoT helps the students to increase their attention span by facilitating a quick learning process. With this technology, the teachers can easily connect with the students in various ways: structures, visual storytelling, sound and animations and simulation of scientific phenomena. The students can access these things from their smartphones and improve their learning experiences (Chweya et al., 2020). In this well, they grab the learning concept more quickly than ever before within a short time. They can also provide quick feedback on their knowledge by attending various quizzes. Another major advantage of IoT is that it can create an inclusive learning environment for differently-abled students. With this technology, the administration of a learning institution can meet the specific needs of autistic children and other kinds of learning disorders or any kind of physical deformity. Schools can create specific learning programs for them and specially-abled students can access those learning programs on their smartphones (Mishra et al., 2020). For example, the students with impaired vision can choose the option to reach text but they can get their textbook read out to them.

The IoT technology can identify the part of the textbook where the students find it difficult and make changes automatically in the study material accordingly (Satu et al., 2018). This technology is also beneficial for the teachers to create a custom interactive environment for the students, especially for the differently-abled kids. Microsoft has focused on using IoT technologies to transform students into active learners in India. This organization is also researching providing special tools for differently-abled children by making symbol-based communication mobile applications to help with speech disorders. This technology is also useful in the case of automation in the classroom, and it allows the teachers to free up their valuable time and invest it into more meaningful work.

IoT can help the teachers with high-frequency repetitive activities like taking attendance and grading closed-ended questions. With the help of this technology,

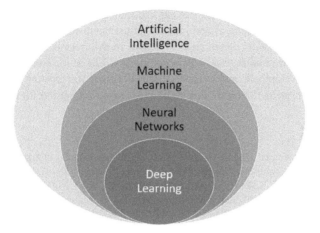

FIGURE 10.3 Integrating artificial intelligence and machine learning with the neural network

the teacher can teach the students in an interactive way to grab the student's attention and make them learn the learning concept accurately. In the smart classroom, the students can transcribe the daily lessons into their smartphones rather than being distracted by taking notes (Satu et al., 2018). The students can save the audio lessons of the teachers and listen to them whenever possible to revise them. With the help of smartphones, the students can access these audio lessons from anywhere at any time. It can be said that IoT can bring a much-needed change in the education industry for better student engagement by providing interactive teaching methods.

Learning is a dynamic, interactive, informative, systematic and scientific process. It is the process of knowledge transfer from instructor to learners. In addition, it is multi-dimensional in nature. The effective teaching-learning process is the blend of audio, visual and kinesthetic tasks promoting the interest of learners and creating a scope of application of knowledge and skills in their personal, professional and public life. It is the physical, social and psychological transformation for the learners. Figure 10.3 is the holistic representation of when AI, ML and neural networks are well integrated leading to deep learning. It is a futuristic perspective. The computer vision, algorithm, network, learning, database management and integration will transform the world of education at different levels such as school, college or university (Webb, Fluck, Magenheim, 2021). It would be possible to provide better quality and affordable education for all achieving the goal of sustainable development and creating win-win situations for various stakeholders of society.

10.4 CONCLUSION AND FUTURE SCOPE

All of the above discussion shows that AI, ML and neural networks can contribute to making the teaching-learning process more effective than ever. In addition, their integration leads to 'deep learning' wherein the learners' behavioral patterns, styles and preferences can be identified and similar inputs can be provided to minimize the demand-supply gap. The future scope exists in utilizing such platforms for talent acquisition, training and development and knowledge management in multinational corporations wherein thousands of employees located in different parts of the world can be connected and trained at ease with better results. The scope can also be extended to leadership development and excellence transforming the existing workforce into human capital meeting the future demands and staying competitive all the time.

REFERENCES

Chen, X., Zou, D., Xie, H., Cheng, G. and Liu, C., 2022. Two decades of artificial intelligence in education. *Educational Technology & Society*, *25*(1), pp. 28–47.

Chweya, R., Ajibade, S.S.M., Buba, A.K. and Samuel, M., 2020. IoT and big data technologies: opportunities and challenges for higher learning. *International Journal of Recent Technology and Engineering (IJRTE)*, *9*(2), pp. 909–913.

Clark, D., 2016. AI – The many ways artificial intelligence will shape teaching & learning. Presented at E-Learn: World Conference on E-Learning. Retrieved March 7, 2022 from https://www.learntechlib.org/primary/p/178460/.

Conati, C., Porayska-Pomsta, K. and Mavrikis, M., 2018. AI in education needs interpretable machine learning: lessons from open learner modelling. arXiv preprint arXiv:1807.00154.

Giannakos, M., Voulgari, I., Papavlasopoulou, S., Papamitsiou, Z. and Yannakakis, G., 2020. Games for artificial intelligence and machine learning education: Review and perspectives. In M. Giannakos (Ed.), *Non-Formal and Informal Science Learning in the ICT Era* (pp. 117–133). Springer: Singapore. DOI: https://doi.org/10.1007/978-981-15-6747-6

Guan, C., Mou, J. and Jiang, Z., 2020. Artificial intelligence innovation in education: a twenty-year data-driven historical analysis. *International Journal of Innovation Studies*, *4*(4), pp. 134–147.

Haristiani, N., 2019. Artificial intelligence (AI) chatbot as language learning medium: an inquiry. *Journal of Physics: Conference Series*, *1387*(1), p. 012020.

Hwang, G.J. and Tu, Y.F., 2021. Roles and research trends of artificial intelligence in mathematics education: a bibliometric mapping analysis and systematic review. *Mathematics*, *9*(6), p. 584.

Iatrellis, O., Savvas, I.K., Kameas, A. and Fitsilis, P., 2020. Integrated learning pathways in higher education: a framework enhanced with machine learning and semantics. *Education and Information Technologies*, *25*(4), pp. 3109–3129.

Luckin, R., 2018. *Machine Learning and Human Intelligence: The Future of Education for the 21st Century*. UCL IOE Press, UCL Institute of Education, University of London, London, UK.

Maksimović, M., 2018. IOT concept application in educational sector using collaboration. *Facta Universitatis, Series: Teaching, Learning and Teacher Education*, *1*(2), pp. 137–150.

Malik, G., Tayal, D.K. and Vij, S., 2019. An analysis of the role of artificial intelligence in education and teaching. In P.K. Sa, S. Bakshi and M.N. Sahoo (Eds.), *Recent Findings in Intelligent Computing Techniques* (pp. 407–417). Springer, Singapore.

Mishra, A.S., Karthikeyan, J., Barman, B. and Veettil, R.P., 2020. Review on IoT in enhancing efficiency among higher education institutions. *Journal of Critical Reviews*, *7*(1), pp. 567–570.

Ramlowat, D.D. and Pattanayak, B.K., 2019. Exploring the internet of things (IoT) in education: a review. In S.C. Satapathy (Eds.), *Information Systems Design and Intelligent Applications* (pp. 245–255). Springer: India.

Satu, M.S., Roy, S., Akhter, F. and Whaiduzzaman, M., 2018. IoLT: an IOT based collaborative blended learning platform in higher education. In *2018 International Conference on Innovation in Engineering and technology (ICIET)* (pp. 1–6). IEEE.

Shah, D., Patel, D., Adesara, J., Hingu, P. and Shah, M., 2021. Exploiting the capabilities of blockchain and machine learning in education. *Augmented Human Research*, *6*(1), pp. 1–14.

Temitayo Sanusi, I., 2021. Teaching machine learning in K-12 education. In *Proceedings of the 17th ACM Conference on International Computing Education Research* (pp. 395–397).

Webb, M.E., Fluck, A., and Magenheim, J. (2021). Machine learning for human learners: opportunities, issues, tensions and threats. *Education Tech Research Dev*, *69*, pp. 2109–2130. Available at: https://doi.org/10.1007/s11423-020-09858-2

Zawacki-Richter, O., Marín, V.I., Bond, M. and Gouverneur, F., 2019. Systematic review of research on artificial intelligence applications in higher education–where are the educators? *International Journal of Educational Technology in Higher Education*, *16*(1), pp. 1–27.

11 IoT-Based Smart Mask to Combat COVID-19

Harveen Kaur
G.S.S.D.G.S. Khalsa College

Harpreet Kaur and Navjot Kaur
Punjabi University

CONTENTS

11.1 INTRODUCTION

Novel Coronavirus (COVID-19 or 2019-nCoV), came to the surface in late 2019. It is believed that COVID-19 emerged from Wuhan, the Hubei Province of China. It was first derived from bats and soon got transmitted to humans via raccoon dog being the intermediate host. The significant symptoms of SARS-CoV-2 are fever, shortness of breath, and cough [1]. In the past few decades, the world has faced many pandemics such as Ebola, Spanish Flu but soon they got treated with medical aid. Despite precautious measures taken by the government, COVID-19has led to and is also leading to a lot many deaths across the world. With the rise in cases, the burden on the medical system is also increasing. IoT can prove to be a savior as it can help in keeping a check on the necessary precautionary measures to combat COVID-19, such as maintaining a necessary distance of 6ft., wearing a mask, as wearing a mask reduces the transmission of infection, thereby stopping the chain. Likewise, there are many IoTs such as smart bands, smart watches, smart mobile phone applications, and smart masks which help in monitoring cough frequency and intensity and other COVID-19 related symptoms at just a click away; this all will no doubt release the burden on healthcare systems [2]. All those IoTs that are applicable in the medical field turn out to be the

DOI: 10.1201/9781003250357-11

Internet of Medical Things (IoMT). IoMT is an extension of IoT in the medical field. All those IoTs which when used in diagnosing, treating, and monitoring patient scan be marked under IoMT. IoT devices are not only bound to be a smart watch or smart band but to smart masks and smart drones also, which can help in combating the spread of this pandemic. Smart masks, though not so popular IoMT device, can be efficiently used as a means to early detect, monitor, and therefore combat COVID-19. The best thing about smart masks is that they are the basic mask that is worn by all of us, be it, children, at their school premises or be it any of us at our work premises. The best thing that could ever be incorporated in "Just Basic Smart Masks" is the smartness to work smart, smart enough to help healthcare in these rough times of pandemic. The pandemics which once came to the world never left the world. The history of pandemics proves that pandemics might turn in to an epidemic or endemic but they never left the world. And so, it has become the need to turn the basic masks into smart masks which are smart enough to measure the basic symptom of COVID-19, i.e., cough through its frequency and intensity and generate an alert when the nearby air is contaminated or when airborne germs exist in germs. Masks have become an important attire in our daily life. Wearing masks has become the only means to stop the spread of infection. "If you wish to stop the spread of infection, Wear a Mask" is what we get to hear all the time. To make it a compulsion, a check can be laid by a smart device called smart drones, which figure out from a distance whether the people are wearing a mask or not, and if they are not obeying the rules, their information can be passed to the government officials to make them pay a heavy fine.

11.2 LITERATURE REVIEW

This section discusses the research work done by various researchers related to the role of IoT in diagnosing, monitoring, and combating COVID-19.

Mohammad Nasajpour et al. in their paper Internet of Things for Current COVID-19 and Future Pandemics: An Exploratory Study have explored various applications of IoT in providing a helping hand to healthcare systems. IoT is efficient in early detection and during the treatment phase as well. IoT was proven to help healthcare officials in early diagnosing and treating patients [1].

A Gasmi et al. in their paper have discussed the applications of IoT-enabled devices in general and specific to the medical field. In the medical field, he has researched that IoMT plays a major role in collecting the sample, monitoring patients, treating patients, reviewing contact tracing, and computing the results with edge computing. As a result of testing the validation of IoMT for detecting COVID-19, it was concluded that they play a major role in effective treatment, testing, and monitoring of patients [2].

Naren Vikram Raj Masna et al. in their paper have explored various applications of smart mask in providing a helping hand to healthcare systems. Along with protection from germs and infection, a smart mask also can detect the presence of airborne pathogens and can help in protecting humans against the infection with the help of actuators to measure the size of particles in the air [3].

Barnali Ghatak et al. in their paper have proposed a self-powered smart mask for COVID-19. They proposed a three-layer facial mask. The best part about them is that they do not need an external power supply but they become active with breathing cycles. Along with being energy efficient, they protect humans from catching an infection from airborne germs due to their triple layer [3].

Rohan Reddy Kalavakonda et al. in their paper proposed a smart mask as a protection against airborne pathogens. The active mask proposed is efficient enough to sense the size and quality of particles in the air and then if they can cause a threat, the mist is sprayed to destroy the particles. It also generated real-time alerts to recharge, refill, and decontaminate before reusing it [4].

Noori Kim et al. in their paper proposed a customized smart mask for healthcare personnel. The mask is customized for healthcare personnel depending upon their facial features and requirements. This research is based on the concept that each human being has different facial features and so one size and shape of mask can't comfortably fit all. And comfort matters, especially in the case of healthcare personnel as they have to wear it the whole day long [5].

Liang Zhu et al. in their paper proposed a smart face mask for cough monitoring with harmonic frequency detection. The wireless harmonic tag is attached for real-time cough monitoring and alert in case the cough frequency or intensity is beyond the threshold. The mechanism used by them is Clutter Rejection. They figured out that lightweight and zero power devices will be beneficial for monitoring cough intensity and frequency. The cough events are monitored with the Received Signal Strength Indicator (RSSI) [6].

Andrea Fois et al. proposed an innovative smart face mask to protect healthcare workers from COVID-19 infection as they are the leaders in helping the world fight pandemic and their safety is of utmost importance. AG-47 smart mask is proposed for continuous monitoring of temperature inside the mask, humidity, heart rate, and oxygen level in haemoglobin [7].

Lalitha Ramadass et al. proposed the deep learning technique to lay down the check on social distance in public places with the help of drones. The YOLOV3 algorithm is embedded in the camera of drones to detect the distance between the people in public places. The accuracy of 0.95 is obtained and the drones are programmed to generate an alert if the distance is not maintained [8].

Faris A. Almalki et al. in their paper studied the detection of the mask on the face with drones to assure the mask is worn by people. For the face, detection drones use Convolution Neural Network (CNN) with Modified Artificial Neural Network (MANN). Promising accuracy of 82.63% is obtained in face detection. With the value of 0.98 for F1 score 0.98, this paper aims at reducing the COVID spread by assuring the check on the mask [9].

11.3 ROLE OF IoMT DEVICES IN REAL LIFE

This section discusses the role of various IoMT devices in detecting and combating the spread of COVID-19.

The term Internet of Things (IoT) was coined by Kevin Ashton. IoTs gained popularity due to their smart technology of connecting all smart objects within a network and that also without human intervention [8]. Generally, IoT can be defined as the devices that can connect to the internet for transferring and monitoring data.

IoMT has offered a relieving helping hand to the healthcare systems.

Though the symptoms of COVID-19 (i.e. fever, common cold, cough and fatigue) are general, yet they can prove to be deadly if not properly diagnosed at an early stage. The incubation period of COVID-19 is of around 14 days. An asymptotic patient who is in his incubation period is not aware of the COVID virus residing inside his body can be the transmitter of COVID-19 virus and this nature of the virus has hit the healthcare system so badly; because if the patient is not quarantined in his early stage of incubation, then he can form a chain and this endless chain is the reason why the world is facing the second wave of COVID-19 [10,19].

On one hand, clinical approaches are being laid down by medical researchers to mitigate this disease, and, on the other hand, non-clinical approaches have also actively come forward to break this chain. And out of those non-clinical approaches, IoMT technology came up to be the one that is efficient in tracking, diagnosing, and monitoring a patient.

IoMT has offered a helping hand to the leading fighters, i.e., healthcare officials in their tough times. Generally, they can be categorized as Wearables and Non-wearables. Smart drones are non-wearable, whereas smart masks are wearable IoMT. They are discussed as follows.

11.3.1   WEARABLE IoMT

Wearable IoMTs are the electronics that can be worn. As per the research done by Juniper [20], wearable IoMT can be defined as computing technologies that are application-based and are capable of receiving and processing the output readings when made to be worn or simply stick to the body such as smart bands, smart glasses, smart thermometers, smart masks, and many more. Mask has already become a necessary part of our daily attire. It is one of the most important precautionary measures that can help in stopping the spread. Smart mask, an important wearable IoMT, is discussed below.

- **Smart Mask,** the basic mask with the ability to act smart, smart enough to test the quality of air and therefore generating an alert when the quality is poor and also helps in measuring the intensity and frequency of cough, thereby generating an alert when the person is coughing frequently and heavily. Masquerade masks are the smart masks with smartness to combat COVID-19. Be it a smart mask or a basic mask, it is a must to wear it and therefore we need a check to assure it is worn in public places. Figure 11.1 describes the smart mask module in collecting data related to location, droplet size, and range.

11.3.2   NON-WEARABLE IoMT

Non-wearable IoMTs are those devices that cannot be worn but are capable of processing input given through programmed devices. Non-wearables are capable of replacing humans in sanitizing, treating, and diagnosing the patient at an early stage which is way too important to mitigate the spread of the pandemic. Robots, smart

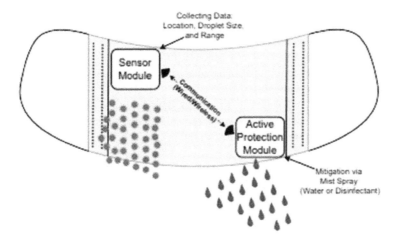

Collecting Data:
Location, Droplet Size,
and Range

Sensor
Module

Communication
(Wired/Wireless)

Active
Protection
Module

Mitigation via
Mist Spray
(Water or Disinfectant)

FIGURE 11.1   Smart mask module collecting data [4]

helmets, and smart drones are a few examples of non-wearable IoMTs. Drones, the example of non-wearable devices, are briefly discussed below.

- **Drones:** Drones are the ones that fly with or without human intervention. Drones are also named Unmanned Aerial vehicles (UAVs) which may or may not need to be controlled remotely by human beings, but they need sensors, Global Positioning System (GPS), and other communication services to build up a connection. Internet of Drones Things is the name given to the implantation of IoTs in drones [11]. By doing so, the drone becomes capable of monitoring, searching, and delivering. Drones can, no doubt, reach a place that otherwise might be hard to reach. Examples of various IoT drones are medical drones, surveillance drones, disinfectant drones, thermal imaging drones, etc. Given below are the examples of various IoT-based drones:
  - **Thermal Imaging Drones** are capable of capturing the temperature of humans in a crowd and notifying others in connection with it so that distance can be kept and, as a result, leading to fewer human interactions [7].
  - **Disinfectant Drones** as the name suggest are the ones that are used to disinfect the contaminated zones, therefore, reducing the risk of human workers getting infected during the process of disinfecting/sanitizing the contaminated zone [11].
  - **Medical Delivery Drones,** as the name suggests, are responsible for delivering the medical needs to the patients, therefore, reducing the need for patients to visit hospitals for medical needs. With such services, people will have quick accessibility to treatment [11].
  - **Surveillance Drones** are necessary to curb this increasing pace of COVID-19 by monitoring social gatherings and assuring that people

are following the guidelines given by healthcare systems to maintain a social distance of 6 ft., wear masks, and do not gather in a group [11].

- **Announcement Drones,** as the name suggests, are drones meant for broadcasting any kind of important information regarding COVID-19 even in the containment areas where it might be a risk to send workers for announcement [11].
- **Multipurpose Drones** are drones for doing multiple kinds of tasks such as sanitizing areas, monitoring crowds, broadcasting the latest information, monitoring temperature, and many of such sorts [11].

The major contribution done to the medical system by drones is in monitoring socially, whether people are wearing a mask or not, and generating an alert if the guidelines laid by the government are not adhered to.

And so drones and smart masks are truly a savior in these tough times. From the literature, the analysis of drones and smart masks based on their effectiveness and application is shown in Table 11.1. Various parameters covered by them are figured out in Table 11.2.

11.4 NEED FOR SMART MASK AND DRONE-ASSISTED CHECK ON SMART MASK

11.4.1 Need

To bring an end to this infamous pandemic, IoMT has come to the surface. Besides other IoMT devices, not so famous and not so common smart mask could be a savior in these tough times when the people had already lost their lives to this pandemic and the world cannot afford more.

TABLE 11.1

Comparative Analysis of Drone and Smart Mask in Early Diagnosing COVID-19

IoMT Device	Application	Capability	Challenges
Smart Mask [4]	• Location tracing • Cough intensity • Cough frequency • Detecting airborne pathogens • Protecting against the transfer of COVID-19 germs	• Early diagnose the COVID-19 infection by measuring the intensity and frequency of cough	• Difficult to wear whole day long
Drone [11]	• Provide a contactless sample collection and delivery of medicine to infected patients • Monitor Wearing of Mask and generate an alert	• Reduced risk to healthcare officials of getting infected • Help reduce the spread by keeping a check on mask	• The battery discharge rate of drones

TABLE 11.2
Parameters Covered Corresponding to Smart Mask and Drones

IoMT Device	Parameters
Smart Mask [4]	• Cough intensity
	• Cough frequency
	• Location
	• Temperature monitoring
	• Amount of droplets
	• Presence of airborne pathogens
	• Contaminated air
Drones [11]	• Sample collection
	• Check on mask
	• Generate alert
	• Medicine delivery
	• Disinfecting

Smart mask is capable to generate alerts when a person with a symptom such as a heavy cough is detected, and in addition to this masks could also trace the location so that if he is been detected positive his first contact persons can be informed to isolate themselves and get tested.

Smart masks are already so commonly worn by us all as "Just Masks". Turning these so common masks to smart masks will help in combating COVID-19. Be it schools, colleges, or offices, wearing a smart mask is already compulsory and has become a part of our attire. Easy to handle, manage, and even school-going students, which due to lack of understandability are more prone to the risk of getting infected, wear masks on daily basis. Turning these masks smart has and will help technology save the lives of humans of all age groups. Smart mask, no doubt, has been made compulsory, but still people skip wearing it due to many reasons due to which the pandemic has not come to an end yet; therefore, a check has been laid to make sure people wear a mask, and if they fail to wear the drones will trace the identity of person and generate an alert to notify the government officials and a heavy fine will be laid down.

11.5 RESEARCH OBJECTIVES AND METHODOLOGY

The objective of the study is to analyze various existing IoMT techniques to detect COVID-19 and to propose a smart mask with advanced features to early detect and combat COVID-19 and also to study the drone-assisted smart masks to assure people wear the mask in public places so that chain can be brought to an end.

11.5.1 METHODOLOGY

The proposed work will be carried out in various phases. Smart mask is incorporated with sensors for marking out the body temperature, Respiration Rate (RR), Location

Tracker (Loc) of a person in range, cough intensity, and frequency of cough. Along with these features, smart mask can be linked with smartphones to generate an alert for a person when he is not wearing a mask while stepping out of the house, making sure the HOME location is saved in the phone to assure the phone does not generate an alert to wear a mask while the person is at home. No doubt, masks have become part of our attire, but we do forget to wear them sometimes; and if smart mask generates an alert on our phone notifying us to wear a mask, that could truly be a savior. Moreover, with an incorporated feature of face detection at some proximity, smart masks would be able to notify when a person not wearing mask is close As it has been researched that a person 'Not Wearing' mask is more prone to risk of getting infected, than the one 'Wearing' mask. Figure 11.2 depicts the check on mask with a smart mask.

And if anyone tries to cross 6 ft. of the required distance or if someone without wearing a mask is approaching, then an alert could help us in maintaining the required distance of 6 ft. Figure 11.3 depicts the check on distance with smart masks.

A smart mask triggers a signal to identify if any person in a certain range is not wearing a mask or has any of the COVID symptoms. A signal is triggered and

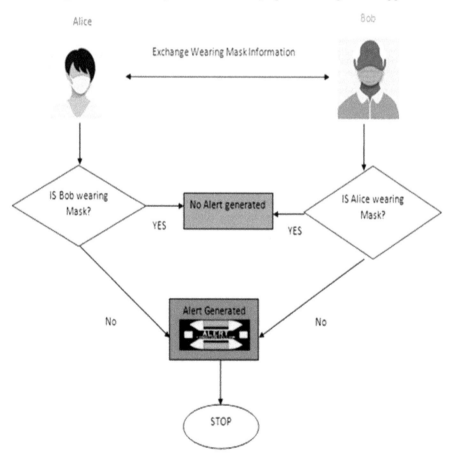

FIGURE 11.2   The check on "Mask" with smart masks

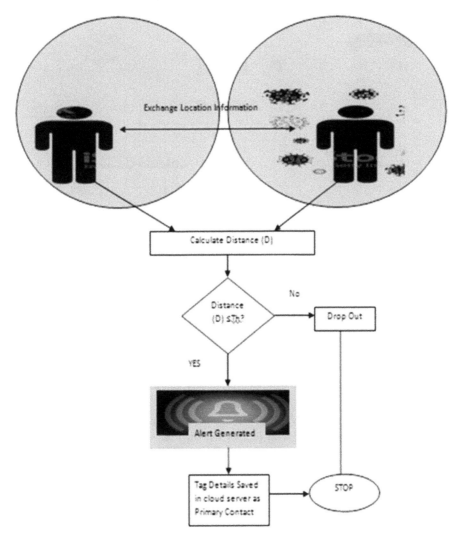

FIGURE 11.3 The check on distance with smart masks

when a person comes in the range, the information is exchanged between them. The information exchanged is medical readings (body temperature, wearing a mask(-yes/no), distance, RR, cough intensity, cough frequency, personal information, home location, active GPS location, nearby healthcare official's contact details); with the exchange of this information, sensors actively trace out the values of medical readings and compare them with the respective threshold. And if any of the checks are true, then a vibration alert is generated by which the person gets alerted about the suspected one. With the alert, the information of the suspected patient is passed onto the cloud and to the linked officials. Figure 11.4 depicts the flow of smart masks in checking the parameters.

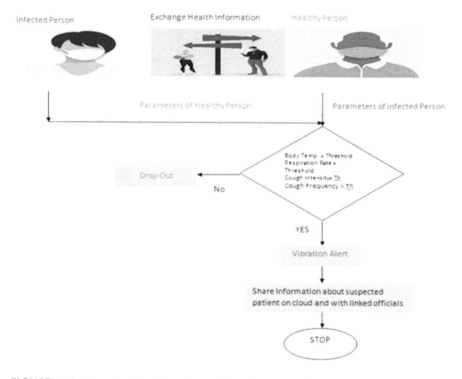

FIGURE 11.4   Flowchart depicting the working of smart masks

Drones fly in the air and have been made smart enough to detect the face of a person or the persons in the group to check if they are wearing a mask or not. If they wear a mask, no action takes place; but if the mask is not worn and an alert is generated, the government officials are informed and a heavy penalty is laid down. Figure 11.5 depicts the flow of check on masks by smart drones.

11.6   CONCLUSION

The infamous COVID-19 pandemic has caused more than a million deaths across the world. To help stop the chain, IoMTs have come to the surface. IoT does play a vital role in stopping the spread by tracing the locations, monitoring the temperature of the people through thermal scanning, assuring the healthy distance of 6 ft. is maintained, and assuring that people wear a mask. From the gist of the paper, it can be figured out that thermal sensors, when incorporated into masks, are effective enough for testing the body temperature of a person from a particular distance and generate a notifying alarm on encountering a person with a high body temperature in the crowd and also notify when the precautionary distance of 6 ft. is not maintained. Smart masks can be widely used even in schools, offices, and rural areas as they don't need any expert knowledge to deal with smart masks. Drone-assisted checks to assure people wear a mask in public places will surely help in combating the spread of this pandemic. Ranging from thermal scans done by smart masks to a check on the smart mask with

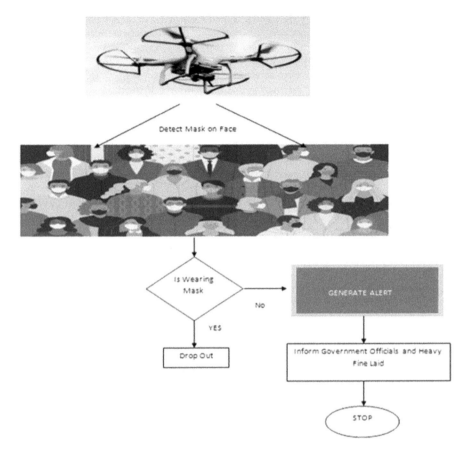

FIGURE 11.5 Drone-assisted check on smart mask

smart drones, IoMTs play a vital role and can surely prove to be effective in diagnosing, monitoring, and combating COVID-19.

REFERENCES

1. Nasajpour, M., Pouriyeh, S., Parizi, R.M., Dorodchi, M., Valero, M. and Arabnia, H.R., 2020. Internet of Things for current COVID-19 and future pandemics: An exploratory study. *Journal of Healthcare Informatics Research*, pp.1–40.
2. Gasmi, A., 2021. Enabled IoT applications for Covid-19. In S. Kautish, S.-L. Peng, and A.J. Obaid (Eds.), *Computational Intelligence Techniques for Combating COVID-19* (pp. 305–331). Springer, Cham.
3. Kalavakonda, R.R., Masna, N.V.R., Bhuniaroy, A., Mandal, S. and Bhunia, S., 2020. A smart mask for active defense against coronaviruses and other airborne pathogens. *IEEE Consumer Electronics Magazine*, 10(2), pp.72–79.
4. Kim, N., Wei, J.L.J., Ying, J., Zhang, H., Moon, S.K. and Choi, J., 2020. A customized smart medical mask for healthcare personnel. In *2020 IEEE International Conference on Industrial Engineering and Engineering Management (IEEM)* (pp. 581–585). IEEE.

5. Zhu, L., Ha, T.D., Chen, Y.H., Huang, H. and Chen, P.Y., 2022. A passive smart face mask for wireless cough monitoring: A harmonic detection scheme with clutter rejection. *IEEE Transactions on Biomedical Circuits and Systems.* 16(1), pp. 129–137.

6. Fois, A., Tocco, F., Dell'Osa, A., Melis, L., Bertelli, U., Concu, A., Bertetto, A.M. and Serra, C., 2021. Innovative smart face mask to protect workers from COVID-19 infection. In *2021 IEEE International Symposium on Medical Measurements and Applications (MeMeA)* (pp. 1–6). IEEE.

7. Ramadass, L., Arunachalam, S. and Sagayasree, Z., 2020. Applying deep learning algorithm to maintain social distance in public place through drone technology. *International Journal of Pervasive Computing and Communications,* 16(3), pp. 223–234.

8. Almalki, F.A., Alotaibi, A.A. and Angelides, M.C., 2021. Coupling multifunction drones with AI in the fight against the coronavirus pandemic. *Computing,* 104, pp. 1033–1059.

9. Rahman, M.M., Manik, M.M.H., Islam, M.M., Mahmud, S. and Kim, J.H., 2020. An automated system to limit COVID-19 using facial mask detection in smart city network. In *2020 IEEE International IOT, Electronics and Mechatronics Conference (IEMTRONICS)* (pp. 1–5). IEEE.

10. Khan, H., Kushwah, K.K., Singh, S., Urkude, H., Maurya, M.R. and Sadasivuni, K.K., 2021. Smart technologies driven approaches to tackle COVID-19 pandemic: a review. *3 Biotech,* 11(2), pp.1–22.

11. Elsayed, E.K., Alsayed, A.M., Salama, O.M., Alnour, A.M. and Mohammed, H.A., 2020. Deep learning for Covid-19 Facemask detection using autonomous drone based on IoT. In *2020 International Conference on Computer, Control, Electrical, and Electronics Engineering (ICCCEEE)* (pp. 1–5). IEEE.

12. Rehman, A., Sadad, T., Saba, T., Hussain, A. and Tariq, U., 2021. Real-time diagnosis system of COVID-19 using X-ray images and deep learning. *It Professional,* 23(4), pp.57–62.

13. Ahuja, S., Panigrahi, B.K., Dey, N., Rajinikanth, V. and Gandhi, T.K., 2021. Deep transfer learning-based automated detection of COVID-19 from lung CT scan slices. *Applied Intelligence,* 51(1), pp.571–585.

14. Singh, R.P., Javaid, M., Haleem, A., Vaishya, R. and Ali, S.,2020. Internet of Medical Things (IoMT) for orthopaedic in COVID-19 pandemic: Roles, challenges, and applications. *Journal of Clinical Orthopaedics and Trauma,* 11(4), pp.713–717.

15. Ndiaye, M., Oyewobi, S.S., Abu-Mahfouz, A.M., Hancke, G.P., Kurien, A.M. and Djouani, K.,2020. IoT in the wake of COVID-19: A survey on contributions, challenges and evolution. *IEEE Access,* 8, pp.186821–186839.

16. Garg, L., Chukwu, E., Nasser, N., Chakraborty, C. and Garg, G.,2020. Anonymity preserving IoT-based COVID-19 and other infectious disease contact tracing model. *IEEE Access,* 8, pp. 159402–159414.

17. Dhinakaran, V., Surendran, R., Shree, M.V. and Tripathi, S.L., 2021. Role of modern technologies in treating of COVID-19. In S.L. Tripathi, K. Dhir, D. Ghai, and S. Patil (Eds.), *Health Informatics and Technological Solutions for Coronavirus (COVID-19)* (pp. 145–157). CRC Press.

18. Kumari, S., Ranjith, E., Gujjar, A., Narasimman, S. and Zeelani, H.A.S., 2021. Comparative analysis of deep learning models for COVID-19 detection. *Global Transitions Proceedings,* 2(2), pp.559–565.

19. Dairi, A., Harrou, F., Zeroual, A., Hittawe, M.M. and Sun, Y., 2021. Comparative study of machine learning methods for COVID-19 transmission forecasting. *Journal of Biomedical Informatics,* 118, p.103791.

20. Tatiana, V.S., Osipov, V.S. and Paramonov, A.S., Study of practical implementation of IoMT: creating a web application for distant high-risk group patient's heart-rate tracking.

12 Encryption Algorithms for Cloud Computing and Quantum Blockchain
A Futuristic Technology Roadmap

Prabhsharan Kaur and Isha Sharma
Chandigarh University

Rahul Kumar Singh
University of Petroleum and Energy Studies

CONTENTS

DOI: 10.1201/9781003250357-12

12.1 INTRODUCTION

Cloud computing is emerging technology that follows pay as you use service approach. It provides more flexibility in data availability compared to other approaches. The subsequent sections are focused on detailed discussion in the field of cloud computing.

12.1.1 Cloud Computing

Cloud computing is an emerging technology that uses a network of servers, located remotely and hosted over the internet for storing, processing, and managing data, instead of utilizing personal computers or local servers [1]. In other words, cloud computing is sending services (cloud services) to the other side of the organization's firewall. Services like storage, applications, and other services are fetched through the internet. These services are then delivered through the internet to the users and users pay for the services as per need or pay as per use business model. The term "cloud computing" is strongly associated with IaaS (Information as a Service), SaaS (Software as a Service), and PaaS (Platform as a Service) [2].

12.1.2 Characteristics of Cloud Computing

i.   The cloud computing domain allows the users to access the internet via different web browsers irrespective of their location and time and even when the type of device being used is any computing device that can allow running web browser applications on it.

ii. The cloud services and the infrastructure are directly accessible that a cloud service provider provides via the internet. Due to this, the user cost to buy the resources is significantly reduced as only basic skills are needed for operation.

iii. For the sole purpose of disaster management and business longevity, cloud computing offers reliable services where users can look for multiple web portals to access the computing resources required [3].

iv. The resources provided by cloud computing service providers cost them huge profits as the same will be shared by a large number of users throughout the world.

v.   Because the infrastructure is not owned by the users and they don't have the physical resources provided by the cloud provider at their end, there is the only requirement of maintaining the same at one end that is the service provider end.

vi. Pay per use model also helps the user to check the usage of the services he or she is subscribed to [4].

vii. Scalability is also one of the important features that cloud computing offers.

12.1.3 ARCHITECTURE OF CLOUD COMPUTING

The architectural view of cloud computing divides this technology into two sides: Front side and Back side. The connection between the two sides is the internet layer. The cloud users form the part of the front side that also encompasses the devices they are using to access the data and physical resources available through the internet to the user. The back side relates to the cloud service providers. They have access to a huge number of, though, countable physical resources like computers, memory for backup, network bandwidth, and servers for supporting services [5].

12.1.4 SERVICES OF CLOUD COMPUTING

Cloud computing supports services such as SaaS, PaaS, and IaaS [6]. All these services are utility services as they are subscription-based, i.e., Pay as per use. There are three deployment models as well: Private Cloud, Public Cloud, and Hybrid Cloud, as shown in Figures 12.1 and 12.2.

With cloud storage, saving data in the user's personal system is not required; rather, it can be stored at a remote location. Therefore, preventing unauthorized access to a user's resources and securing resource sharing becomes critical. Malicious attackers can attack service providers providing cloud services, which can result in losing the users' confidential information stored on the cloud. This raises the issue of ensuring the confidentiality of data files [7]. Data storage in an online cloud is termed cloud storage [8], where the data can be accessed and stored from a number of interconnected and dispersed resources that make up a cloud.

Cloud storage can be beneficial in terms of accessibility and reliability, strong protection for backup, rapid deployment, recovery purposes and minimizing storage costs because the organization does not have to purchase and manage expensive hardware, whereas the issues of security and compliance do occur in cloud storage [9].

FIGURE 12.1   Cloud computing services

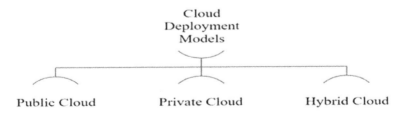

FIGURE 12.2   Cloud computing models

Security is always a major concern in cloud computing since everything from user data to service provider resources are available on the internet and there are always chances of security breaches as well. Due to this, the acceptance of cloud computing technology is still in the infancy stage and it's not fully adopted by all the users throughout the world [10].

This chapter consists of security implications in the second section and then literature survey in the third section providing a review and methodologies. The fourth section with a comparison table compares these methodologies. In the last section, the conclusion and future scope are given.

12.2 SECURITY IMPLICATIONS OF CLOUD COMPUTING

The following main implications may arise in the domain of cloud pertaining to the security subdomain:

   i. Cloud providers in order to provide efficiency in resource utilization decrease the cost. The customers based upon their usage increase or decrease the consumption of resources. The cloud model meets such requirements by delivering key characteristics of elasticity and multitenancy which leads to security implications for cloud models.
   ii. Multitenancy means sharing of resources such as storage, applications, computational resources, and services with other tenants. Different realization approaches are there in multitenancy [11]. The first approach provides dedicated instances with their own customizations for each tenant. The second approach allows each tenant to use their instance but they are the same with different parameters or interfaces. The third approach allows tenants to share the same instance having runtime configurations. The fourth approach allows tenants based upon the load to redirect requests for a suitable instance. Therefore, the third and fourth approaches lead to high risks because the process in hardware and memory are shared among them which violates the confidentiality of the tenant's assets. Location transparency and isolation among the data of tenants is a must to secure multitenancy, which can prevent many attacks [6].
   iii. Scaling up or down the resources based upon recent demands for the services is known as elasticity. The confidentiality issue arises when the scaling up or down tenant's resources allows the next user to use resources of the previously assigned tenant Additionally, the list of allocated resources for services is maintained in a pool which can lead to various threats to security [1].

12.3 LITERATURE SURVEY SUPPORTING SOLUTIONS FOR IMPARTING SECURITY IN CLOUD LAYER

12.3.1 CLOUD COMPUTING SECURITY SOLUTIONS

Common cloud computing solutions include cryptographic approaches. The plain text is encrypted to cipher text using a key known as encryption at one end; and, at the other end, the cipher text is converted back to plain text using a key known as decryption and combining both is cryptography. The key used for both encryption and decryption is the same in symmetric cryptography and a pair of the public key and the private key is used in asymmetric cryptography. The major concern for cloud storage regarding security is using the mechanism to provide secure usage for the users for which various approaches are suggested such as double encryption or splitting the secret key into two parts. Double encryption provides two encryptions on the data to secure it; whereas in splitting a secret key, one part of the key is stored in the system and the other is embedded in a security device [12]. The chapter further explains two-factor mechanisms – attribute-based mechanism and authority-based mechanism – for cloud storage systems in detail.

12.3.2 Two-factor Data Security Protection Mechanism for Cloud Storage System

Through a cloud storage server, the sender sends an encrypted text to the receiver by using this mechanism [13]. Only the identity of the receiver is required by the sender and no more information like a certificate or public key is required. For decryption of cipher text, the receiver needs to know only two things, which are the user's secret key stored in the system and connected to this system a unique security device. If either of mentioned information is missing, the decryption of the cipher text would not be possible. The encryption process is executed two times. Firstly, plain text is encrypted for the identity of the user. After which the text is encrypted related to the public key of the security device. During decryption, once the security device decrypts; then the partially decrypted cipher text is delivered to the system which uses the secret key for decryption. If the user's security device is lost, the encrypted content in the cloud cannot be decoded. Therefore, this mechanism could not support revocability or updating of security devices [14].

At AT&T Labs, this example is implemented where user key k1 is used to encrypt the message and uploads that on cloud server. Now, another user key k2 encrypts the k1which is again uploaded to the server. The user becomes the owner of k2. During recovery of the message, k2 is required by the owner for recovering k1 which finally recovers the message. This dual encryption approach is far better than a single key approach for the encryption of the message. However, this approach has a practical drawback that if the owner uses k2, the message cannot be retrieved. The limitation of the system is due to the lack of revocability for encryption [14]. A dual encryption mechanism and a high level of security are the advantages of this approach, but higher cost and difficulty in revocability affect its efficiency [12].

12.3.3 Attribute-based Access Control with Constant-size Cipher Text in Cloud Computing

- Ciphertext Policy Attribute Based Encryption (CP-ABE) with constant cipher text [15] is proposed in this mechanism which uses fixed-size cipher text and calculates bilinear pairing which enhances data transmission and efficiency of the system and minimizes overhead due to space storage. The burden is reduced and risk on a single authority is minimized by inheritance of authorization provided by a hierarchical access control system. Four polynomial-time algorithms are used in CP-ABE. The data file is firstly encrypted by the data owner using a symmetric key Data Encryption Key (DEK) after which DEK is encrypted by utilizing this mechanism using access control policies. Final cipher text is uploaded by the data owner on the cloud server. A symmetric key is required for a user to decrypt and access the data files. The following are the four sub-algorithms:

- SETUP algorithm requires some initial parameters such as attributes description and security information as an input and generates the Public Key (PK) and Master Key (MK).
- ENCRYPTION algorithm generates Cipher Text (CT) after getting PK, message M, and access structure as an input.
- KEY GENERATION generates Secret Key (SK) after getting MK as an input.
- DECRYPTION algorithm generates message M after getting PK, CT, and SK as an input if the combination of attributes fulfils the access structure associated with the cipher text.

In the cloud, encrypted data files are stored by the data owners to ensure the integrity and confidentiality of the data. Data owners release access control to the file's bacon trolling the description ability of users. The complexity of this scheme will make it difficult to decrypt the larger data files. Hence, data file is encrypted with DEK that is symmetric data encryption and generates corresponding cipher text and then encrypts the key DEK with the above technique. Hence, the user decrypts the cipher text of DEK and cipher text of the file to access the data file [16]. Lower cost, efficiency and less burden on a single authority are the advantages of this approach. Moreover, the risk factor is minimized and attribute-based access control is implemented which helps to make this approach a hierarchical attribute-based system. However, constant size cipher text puts the burden on the system and bilinear pairing evaluation has higher complexity. Text operation for generating constant size cipher is practically difficult to achieve.

Usage of the identity-based time-release technique and distributed hash table increased the efficiency of attribute-based access control encryption for secure access control and sharing [17]. Using the user's attribute, the resource is encrypted; then, this encrypted data is further divided into two parts – encrypted and extracted cipher texts. The identity-based time-release encryption algorithm is used to encrypt the key which is already decrypted and then to create sharable cipher text; extracted cipher text is combined with cipher text and then these are distributed over the distributed hash table network. The cloud server stores the encapsulated cipher texts. The algorithm securely provides proficiency in using the cloud environments.

12.3.4 SHARED AUTHORITY-BASED PRIVACY-PRESERVING AUTHORITY AUTHENTICATION PROTOCOL

To resolve the above privacy issue was the major issue of concern for which Shared Authority based Privacy-preserving Authority (SAPA) [16] with the following main objectives was proposed:

- To enhance the user's privacy-related access request for authentication, an anonymous access request matching mechanism is used to achieve shared access authority.
- Distinguishing proof of another protection challenge in a distributed storage and checking its unobtrusive security issue when a client claims position to cloud server for information sharing. Here the tested solicitation itself can't uncover the client's protection regardless of whether it can acquire the entrance authority [18].
- The user can reliably access its own data field by applying policy attribute access control for cipher text. Proxy re-encryption is adopted to provide authorized data sharing for multiple users temporarily.

SAPA [19] for cloud storage user recognizes authentication and authorization without compromising the private data of the user. After generating the session identifiers, identity tokens are extracted and transmitted to initiate a new session as a query. On receiving the challenge, Master public key is generated by randomly choosing the secret key and the public key. The access authority policy of user is generated by using set of generated cipher text values. Anonymity of data and user policies are provided using this approach where the hash function helps not to protect data from unauthorized access even if inverse operations are to be followed. Attribute-based access request provides more security and flexibility making it feasible to work with untrusted cloud server, whereas higher complexity and space overhead makes it difficult to apply [20].

12.3.5 HYBRID AUTHENTICATION PROTOCOL

In this mechanism, various approaches of encryption algorithms are combined to achieve security and flexibility. The policy of access control will depend upon quantity of attributes but there will be a change in the size of cipher text. For attribute-based access control, public key is used. Further, master key generation is also required. Therefore, using attributes of data file, unique master key will be generated while encrypting [21].

Further, in this mechanism, attribute-based encryption and identity-based encryption have been merged for secret key generation that will serve as a private key for the user. Sixteen-bit format of a key will make it hard for a hacker to crack encrypted data. Once the encrypted file is stored at the defined path, the original file will also get saved, for backup. Nandgaonkar and Kulkarni [21] has explained the working of hybrid authentication mechanism in his work.

12.4 COMPARATIVE INVESTIGATION OF VARIOUS ENCRYPTION ALGORITHMS SUPPORTING SECURITY AT CLOUD LAYER

Table 12.1 compares different encryption mechanisms, i.e., privacy-preserving, two-factor data security protection, access control using constant text attribute and hybrid authentication mechanism based upon the parameters such as complexity, scalability, key size, security level, cost, and overhead which will help to choose the relevant encryption mechanism.

TABLE 12.1

Comparison between Various Encryption Mechanisms

Parameter	Privacy-preserving Protocol (Shared Authority-based)	Two-factor Data Security Protection Mechanism	Access Control Using Constant Cipher Text (Attribute-based)	Hybrid Authentication Mechanism
Scalability	Scalable	Scalable	Scalable	Scalable
Key Size	Variable	Variable	Constant (16 bit)	Constant
Scheme of Working	Data authorization, Preservation, and Proxy re-encryption are achieved.	Identity-based double encryption mechanism is used.	Constant bilinear pairing for constant size cipher text	Re-encryption technique is used.
Access Control Policy	Third-party authorization scheme	Distributed	Hierarchical (inheritance of authorization)	Hierarchical
Security Level	High	Very high	Secured	High
Revocation Method	Not implemented	Allows revocability of device	Not implemented	Not implemented
Overhead	Complexity overhead	More time to recognize security device	Less space storage	Overhead due to space
Computing Complexity	More	Less	Polynomial	Less
Cost	More	Moderate	Less	Less

After analyzing the different approaches, we can apply them in the domain of cloud computing security and privacy as also given above in Table 12.1. The hybrid authentication encryption algorithm was found to be the optimal approach to solving the highlighted issue of supporting encryption algorithms in the cloud.

The following section will be describing the algorithm using hybrid approach:

Algorithm HAM: Hybrid Authentication Mechanism

Step 1: Start

Step 2: The file is uploaded by the cloud user onto the cloud server.

Step 3: On the cloud server's side, the attributes of the file are generated, such as size of the file say SF, name of the file say NF, and the modification time of the file say MTF.

Step 4: The identity-based encryption is applied and the public key is generated based on it say KEY(PIBE).

KEY(PIBE) = generated using prime number generator G and hash function H and is public in access

Step 5: The secret key is generated say SKey from KEY(PIBE).

[random number *ran* from the set of random values and H]

Step 6: Using the attribute access control technique, the master key is generated say Mkey.

[random exponent a €(ran, H) and generator is used to create Mkey, i.e., secret to all]

Step 7: Verify the secret key, i.e., Skey. If verified, go to step 8; otherwise, go to Step 9.

Step 8: The data is encrypted successfully. The encrypted file and the original file both are stored and then go to Step 10.

Step 9: DISPLAY "Encryption was not successful. Try again" and Exit.

Step 10: End

12.5 QUANTUM COMPUTING AND BLOCKCHAIN

12.5.1 BASICS OF QUANTUM COMPUTING

Quantum computing is a relatively fresh idea of computing that has the potential to transform the industry. Quantum computation is based on the concept of creating computers that are many times more potent than even supercomputers. As we delve deeper into the workings of quantum computers, we'll be confronted with an increasing number of quantum physics principles, the application of which will look more like science fiction than reality. As a result, the eminent physicist Niels Bohr correctly stated, "anyone who is not shocked by quantum theory has not properly understood it" [22,23].

The qubit is the basic building entity particle physics (or quantum bit). Quantum computers, in contrast to ordinary computers, function by manipulating qubits rather than binary numbers or bits. A bit can have either 0 or 1 as its value.

In contrast, a qubit can be either one or the other, or a combination of the two, as shown in Figure 12.3. The uncertainty inherent in quantum mechanics causes this superposition of possibilities. A qubit is a scalar running from the source to a location on the surface of the enclosing sphere [24]. The numbers 1 and 0 represent the north and south poles, respectively. A quantum superposition of 1 and 0 makes up the remaining positions. A quantum computer with more than one qubit uses quantum entanglement, a quantum physics concept.

Entanglement, according to Albert Einstein, is "spooky action occurring at a distance." It will allow one qubit state to get connected with another state. Therefore, changing one qubit's value (with 1 or 0) changes the value of the other. When a qubit is added to a quantum computer, its processing power doubles. If a quantum computer has 1,000 qubits, it can simultaneously test $2^{1,000}$ different input combinations. This is quantum computers' amazing power, which will determine the future of technology [22].

Classical Bit means BIT 0 or BIT 1

Qubit means   BIT 0 or BIT 1 or

COMBINATIONS€ |0> +| 1> / √2

FIGURE 12.3 Basic qubit configuration

12.5.2 How Quantum Fits in Cloud: As-A-Service Quantum Computing

As the cloud computing domain grows, it is confronted with a major challenge that stems from its very success. This is a security problem. Cloud computing will not be totally secure until this problem is completely eliminated [25]. The issue is figuring out how to secure the cloud utilizing present approaches in order to boost corporation and individual trust in cloud computing. The solution to this issue can be found in yet another branch of computing that is on its way from the lab to the real world. Quantum computing has progressed to the next level.

The quantum cloud pair will be a one-of-a-kind technology. It would give cloud computing a new feature called "quantum computing as a service." (Figure 12.4). It would provide such significant technological improvement in the realm of computing that current computers would appear to be obsolete. In the near future, this technological duo will give numerous benefits. It would have the fundamental benefits of combining quantum computing and cloud computing, such as lower overall costs, increased resource sharing, web storage and processing, simplified maintenance, and instant access, flexibility, and mobility.

This technology would result in unforeseeable improvements in the technologies we use every day. The elimination of security problems connected with cloud computing would have been the most significant benefit. This security would be provided via quantum cryptography, which would be an intrinsic characteristic of quantum computing. Quantum cryptography is a technology that uses quantum mechanics to provide information security. The inviolability of quantum physics laws provides the foundation for information security.

Quantum computers would use extremely strong encryption techniques to safeguard data stored in the cloud, which a hacker's machine would have to decode for long periods of time [26]. This would aid in the consumer's acceptance of cloud computing technologies. Customers would also benefit from the quantum cloud duo's tremendous computing power. Quantum computing's super-fast technologies will be accessible via the internet from everywhere. This technology would pave the way for

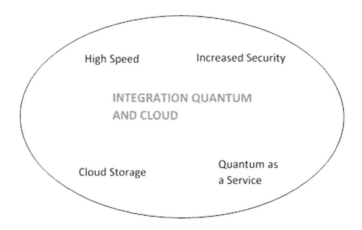

FIGURE 12.4 Quantum integrated cloud computing model

the next generation of computing, making it more portable and perhaps leading to massive productivity increases.

In the sphere of smart portable devices, there have been numerous advancements. High-quality apps would be achievable since the application support infrastructure would be housed on a cutting-edge, shared cloud platform in line with current trends, but at a faster rate than ever before. It would also result in portable gadgets that are slimmer and more powerful. Cloud storage is a type of online storage. This would be yet another benefit for them. Furthermore, such cutting-edge technology will enable organizations, large and small, to embrace and contrivance cloud computing.

It would pave the path for a boost in universal cloud computing adoption, allowing people to experience the computer world in new ways. As a result, cloud computing and quantum computing are important aspects of the computing future. On the other hand, coupling these technologies will lay the groundwork for yet another significant invention, one that will shift computers in a new direction.

Quantum computing's function in cloud computing will also serve as a foundation for future cloud super-fast database operations. Lov Grover spoke about this aspect of quantum technology. Lov Grover developed a quantum approach in 1996 that allowed him to search a quantum repository with P terms in O (root P) steps. Finding the required entry would take P/2 steps in a typical calculation. For a database of 40,000 names, Grover's method would necessitate root of 40,000 which is equal to 400 steps rather than 4,000 for a database of 40,000 names. The method works by superimposing all 40,000 entries into a superposition in which each entry appears to have the same probability of appearing in proportion to the system's evaluation. A quantum computer, based on Grover's notion, might theoretically reduce search time from hundreds of years to minutes [23,27].

As a result, it will be quite useful in the future for providing rapid database operations. As a result, once the quantum computer is complete, our existing *N*-bit cryptography system will be insecure. Local authorities, banks, security businesses, and the military all need to safeguard their data from hackers, as quantum technology may one day be able to decode their systems. As a result, a new secure communication system with strong encryption is required. Quantum science for speed and cryptology for protection are both on the horizon.

Consider the following scenario: two users (Alice and Lara) on the cloud are exchanging and receiving quantum keys via quantum cryptography encryption, and a hacker is attempting to overhear it.

Example

In the cloud, two users (Alice and Lara) send and receive keys using cryptographic encryption technology simultaneously.

At the same time, hackers are attempting to intercept it. Alice being a sender starts by making a basic qubit and sending it to Lara. Alice uses both a diagonal (X) and a rectilinear filter. The symbol used to denote sending of the key is the plus symbol (+). "/", denotes a 45° diagonal. "−" denotes photon spin, while "|" denotes 90° rectilinear spin. Alice analyzes the value of the key using polarization. Lara waits for photons to arrive before applying any rectilinear or diagonal polarization

TABLE 12.2

Sending and Receiving of Keys Using Cryptographic Encryption Technology and Qubits

Alice's polarization	X	+	X	+	X	+	+	+
Alice's spin	\	-	/	/	-	/	-		
Alice's value	0	0	1	0	0	0	0	1
Alice's answers	YES	NO	YES	NO	YES	NO	YES	NO
Lara's polarization	X	+	X	X	+	+	+	X
Lara's spin	\	-	/			\	-	-	/
Lara's value	0	0	1	1	0	0	0	1

filter at random. She maintains track of the polarizing that has been employed, as well as its value and spin. Alice and Lara communicate through an open channel after a successful communication. Lara only gives Alice his polarization filter directives. Alice sends a key [00100001] as shown in Table 12.2 [28].

After that, if Lara receives an improper command of data as given in Table 12.2 [00110001], Lara would be able to determine if the command order is correct or not. The ultimate encoded data can be delivered after the full transmission is completed by correcting the incorrect polarization sequence. In this setup, the eavesdropper is unable to precisely deduce the polarisation sequence. Lara will be able to detect hacker interference in communication if he is unable to decipher the sent data. Other quantum computation criteria, known as quantum non-cloning criteria, ensure that we do not accomplish an indistinguishable clone of the least quantum polarization instance during the calculation. This means that a spy will never be able to reproduce quantum cryptography keys that have been transferred. If someone can clone a state, he can create a large number of identical duplicates of it. He can measure each dynamic variable with random precision at the same moment, thus avoiding the uncertainty principle.

End-user cloud communication is encrypted using the same manner in the quantum internet system. Because quantum internet is based on photon transmission, data delivered via the cloud is instantly transformed if it is intercepted. Finally, once the data has been gathered, the end-user will be able to assess whether or not their discussion was being recorded while it was being conveyed. As a result, building an internet based on quantum mechanics or photons will substantially increase the security medium system. A team of researchers from MIT and Northwestern University has demonstrated long-distance, high-fidelity qubit teleportation [29].

Furthermore, because quantum computation has no replication criteria, we can't have an equivalent clone of any physical state in the middle of the processing. No unauthorized user will be able to acquire a version of the quanta-cryptography-based credentials that are sent. If someone can replicate a state, he can produce a couple of exact replicas. By ignoring the uncertainty principle, he can measure each dynamical variable with arbitrary granularity at the same time. The non-cloning principle, on the other end, eliminates this issue [28].

So, what are the biggest challenges in quantum cryptography? To begin with, the majority of quantum concerns are still theories. Some have been proven, while others are still being tested. Apart from it, the major issues are:

1. It is incredibly difficult to generate a qubit and synchronise numerous qubits at the same time.
2. Because of interference, quantum cryptography transmission capability is too short [30].
3. Quantum dispersed networks, in which we are required to transform manifold qubits into photons depending on the used substance, are the most difficult aspect of quantum cryptography. Long-distance travel necessitates it. Through communication, each and every quantum state can be changed from an atomic system to a photonic system. Because an optical amplifier could damage the qubits, we can't magnify the quantum key carrier signal to send it across long distances. The polarised photon can be corrupted by even a single particle action. It's indeed delicate.
4. Because a photon's rotation may vary when that bounces off other particles, it may not be polarized correctly when it is received.
5. What about the situation when the listener is also in possession of a quantum machine?
6. The operation is too costly, delicate, and has yet to produce results.

12.6   RECENT PROGRESS TOWARDS SOLVING ISSUES

Regardless of the obstacle, except for quantum mechanics, computation progress is still being made. All we need is a powerful quantum computer that can generate numerous qubits, a quantum protocol, and a suitable communication medium.

Google has already stated that its first quantum computer will be based on d-wave techniques. By enhancing the hardware of d-wave, they want to design qubits in a new method. In addition, Google's quantum researchers predict that they and d-wave will develop a new 1,000-qubit processor and make it available to the public [31]. S. Barz et al. demonstrate experimental blind quantum computing for protected cloud computing in a paper published in 2012. They developed the measurement-based quantum computation theoretical framework, which allows a user to express a computation to a quantum server [32].

A group of scientists from Bristol University in the United Kingdom exhibited a quantum-cloud device on September 27, 2013. Qcloud is the name given to this endeavour. The Qcloud quantum computer is housed in Bristol University's quantum photonics facility. The goal is to establish quantum computing as a service in a feasible way Quantum Computing as a Service (QCaaS). Anyone around the globe would be able to access and manage this quantum processor remotely. It would allow users to conduct an experiment and compare the results to their models [33].

Long-distance quantum key transmission necessitates the use of a quantum repeater. We can use photons' "spooky behaviour at a distance" to our advantage. Any information could "teleport" across a vast distance thanks to a link between two pairs of entangled photons. This method might be used to send quantum keys anywhere on the planet. Scientist Anton and his group exhibited a repeater for the first time in 2004 at the Institute of Physics in Vienna. Anton and his colleagues used entanglement to "teleport" data across the Danube at a distance of 600 m, conveyed by a third photon [34].

12.7 CONCLUSIONS AND FUTURE SCOPE

Inevitably, cloud computing is the future of information technology and will be the one-point source of all hardware and software resources. During the implementation and development, there are major concerns that cloud computing should address related to privacy and security before providing the cloud services. Attacks such as data tampering and denial of service are possible in SaaS, whereas data leakage through account and service hijacking is possible in IaaS. Platform as a service encounters malicious machine creations. Because of these security issues, customers are not satisfied with these environments. Therefore, the security algorithms discussed in this paper are providing the basic mechanisms to allow secure usage for the user. For the security, encryption and decryption of data could be done by security algorithms and such algorithms would surely enhance the security framework of any network.

Cloud security being dependent upon computing and trusted cryptography require only authorized user to use their data. The encryption-decryption techniques do not allow the hacker to access the cloud storage. The hybrid algorithm using encryption algorithms provides effective security and storage. They even performed secure computing and are flexible to design. Thus, in conclusion, if security issues are resolved with the use of reliable algorithms, cloud storage for small and large firms would predict a better future.

Cloud computing having significant functionalities for both users and vendors needs to address a huge security gap. Cloud computing is waiting to be adopted after issues related to the security of data as per changing standards and vulnerabilities. Though having a number of mechanisms to protect cloud storage, there is not yet the best mechanism to fully protect the environment through the ever-changing technology and vulnerabilities. For batch auditing encryption and decryption, many advanced algorithms are projected to be developed in the near future. Due to the increase in the need for security, the authentication-based reliable systems are needed to overcome unauthorized access to cloud storage.

REFERENCES

[1] Alouffi, B., Hasnain, M., Alharbi, A., Alosaimi, W., Alyami, H., & Ayaz, M. (2021). A systematic literature review on cloud computing security: threats and mitigation strategies. *IEEE Access*, 9, 57792–57807.
[2] Tabrizchi, H., & Rafsanjani, M. K. (2020). A survey on security challenges in cloud computing: issues, threats, and solutions. *The Journal of Supercomputing*, 76(12), 9493–9532.
[3] Ye, K., & Ng, M. (2018). Intelligent encryption algorithm for cloud computing user behavior feature data. *Journal of Intelligent & Fuzzy Systems*, 35(4), 4309–4317.
[4] Varadharajan, V., & Tupakula, U. (2014). Security as a service model for cloud environment. *IEEE Transactions on Network and Service Management*, 11(1), 60–75.
[5] Das, D. (2018). Secure cloud computing algorithm using homomorphic encryption and multi-party computation. In *2018 International Conference on Information Networking (ICOIN)* (pp. 391–396). IEEE.
[6] Shaikh, A. H., & Meshram, B. B. (2021). Security issues in cloud computing. In *Intelligent Computing and Networking* (pp. 63–77). Springer, Singapore.

[7] Mishra, A., Jain, R., & Durresi, A. (2012). Cloud computing: networking and communication challenges. *IEEE Communications Magazine*, 50(9), 24–25.

[8] Varshini, B., Prem, M. V., & Geethapriya, J. (2007). A review on secure data sharing in cloud computing environment. *Int J Adv Res Comput Eng Technol*, 6(3), 224–228.

[9] Sajay, K. R., Babu, S. S., & Vijayalakshmi, Y. (2019). Enhancing the security of cloud data using hybrid encryption algorithm. *Journal of Ambient Intelligence and Humanized Computing*, 10, 1–10.

[10] Kholidy, H. A. (2021). Detecting impersonation attacks in cloud computing environments using a centric user profiling approach. *Future Generation Computer Systems*, 117, 299–320

[11] Almorsy, M., Grundy, J., & Müller, I. (2016). An analysis of the cloud computing security problem. arXiv preprint arXiv:1609.01107.

[12] Shukla, D. K., Dwivedi, V. K., & Trivedi, M. C. (2021). Encryption algorithm in cloud computing. Materials Today: Proceedings, 37, 1869–1875.

[13] Liu, H., Ning, H., Xiong, Q., & Yang, L. T. (2014). Shared authority based privacy-preserving authentication protocol in cloud computing. *IEEE Transactions on Parallel and Distributed Systems*, 26(1), 241–251.

[14] Liu, J. K., Liang, K., Susilo, W., Liu, J., & Xiang, Y. (2015). Two-factor data security protection mechanism for cloud storage system. *IEEE Transactions on Computers*, 65(-6), 1992–2004.

[15] Teng, W., Yang, G., Xiang, Y., Zhang, T., & Wang, D. (2015). Attribute-based access control with constant-size ciphertext in cloud computing. *IEEE Transactions on Cloud Computing*, 5(4), 617–627.

[16] Yang, K., & Jia, X. (2013). Expressive, efficient, and revocable data access control for multi-authority cloud storage. *IEEE transactions on Parallel and Distributed Systems*, 25(7), 1735–1744.

[17] Namasudra, S. (2019). An improved attribute-based encryption technique towards the data security in cloud computing. *Concurrency and Computation: Practice and Experience*, 31(3), e4364.

[18] Liu, H., Ning, H., Xiong, Q., & Yang, L. T. (2014). Shared authority based privacy-preserving authentication protocol in cloud computing. *IEEE Transactions on Parallel and Distributed Systems*, 26(1), 241–251.

[19] Gomathi, A., & P. Mohanavalli (2015). Anonymous access control by SAPA in cloud computing. *International Journal of Computer Science and Engineering Communications*, 3(2), 848–853.

[20] Li, J., Yan, H., & Zhang, Y. (2020). Identity-based privacy preserving remote data integrity checking for cloud storage. *IEEE Systems Journal*, 15(1), 577–585.

[21] Nandgaonkar, A., & Kulkarni, P. (2016). Encryption algorithm for cloud computing. *International Journal of Computer Science and Information Technologies*, 7(2), 983–989.

[22] Knights, M. (2007). Computing-the art of quantum computing-can'spooky action at a distance'be harnessed to build a new class of computers? *Engineering & Technology*, 2(1), 30–34.

[23] Mullins, J. (2001). The topsy turvy world of quantum computing. *IEEE Spectrum*, 38(-2), 42–49.

[24] Hughes, R. J., & Williams, C. P. (2000). Quantum computing: the final frontier? *IEEE Intelligent Systems and their Applications*, 15(5), 10–18.

[25] Weinhardt, C., Anandasivam, A., Blau, B., Borissov, N., Meinl, T., Michalk, W., & Stößer, J. (2009). Cloud computing–a classification, business models, and research directions. *Business & Information Systems Engineering*, 1(5), 391–399.

[26] Sharbaf, M. S. (2009). Quantum cryptography: a new generation of information technology security system. In *2009 Sixth International Conference on Information Technology: New Generations* (pp. 1644–1648). IEEE.

[27] Hey, T. (1999). Quantum computing: an introduction. *Computing & Control Engineering Journal*, 10(3), 105–112.

[28] Quantiki (2010), The Non-Cloning Theorem. 9 June, 2010. Retrieve from: https://www.quantiki.org/wiki/no-cloning-theorem.

[29] Lloyd, S., Shapiro, J. H., Wong, F. N., Kumar, P., Shahriar, S. M., & Yuen, H. P. (2004). Infrastructure for the quantum Internet. *ACM SIGCOMM Computer Communication Review*, 34(5), 9–20.

[30] Korzh, B., Lim, C. C. W., Houlmann, R., Gisin, N., Li, M. J., Nolan, D., Sanguinetti, B., Thew, R., & Zbinden, H. (2015). Provably secure and practical quantum key distribution over 307 km of optical fibre. *Nature Photonics*, 9(3), 163–168.

[31] "http://www.technologyreview.com/news/530516/google-launches-effort-to-build-its-own-quantum-computer" Web. 3 September 2014.

[32] Barz, S., Kashefi, E., Broadbent, A., Fitzsimons, J. F., Zeilinger, A., & Walther, P. (2012). Demonstration of blind quantum computing. *Science*, 335(6066), 303–308.

[33] Singh, H., & Sachdev, A. (2014). The quantum way of cloud computing. In *2014 International Conference on Reliability Optimization and Information Technology (ICROIT)* (pp. 397–400). IEEE.

[34] Stix, G. (2005). Best-kept secrets. *Scientific American*, 292(1), 78–83.

13 Quantum Artificial Intelligence for the Science of Climate Change

Manmeet Singh
University of Texas at Austin
Indian Institute of Tropical Meteorology

Chirag Dhara
Krea University

Adarsh Kumar
University of Petroleum and Energy Studies

Sukhpal Singh Gill and Steve Uhlig
Queen Mary University of London

CONTENTS

13.1 INTRODUCTION

The Earth's mean temperature has risen steeply over the last few decades precipitating a broad spectrum of global-scale impacts such as glacier melt, sea-level rise and an increasing frequency of weather extremes. These changes have resulted from the rising atmospheric carbon pollution during the industrial era because of the use of fossil fuels. The Earth's mean temperature today is about $1\,°C$ higher than

DOI: 10.1201/9781003250357-13

in pre-industrial times. Recent scientific advances increasingly suggest that exceeding 1.5°C may cause the Earth system to lurch through a cascading set of "tipping points" – states of no return – driving an irreversible shift to a hotter world.

Climate change is global, yet its manifestations and impacts will differ across the planet. Therefore, quantifying future changes at regional and local scales is critical for informed policy formulation. This, however, remains a significant challenge. We begin with a discussion on state-of-the-art science and technology on these questions and their current limitations. With this context, we come to the main theme of this article which is the potential of the emerging paradigm of "quantum computing", and in particular QAI, in providing some of the breakthroughs necessary in climate science.

The rest of the chapter is organized as follows. In Section 13.2, we discuss the science of climate change and the role of artificial intelligence. Section 13.3 discusses QAI for the science of climate change. Section 13.4 concludes the chapter and highlights possible future directions [1,2].

13.2 SCIENCE OF CLIMATE CHANGE AND THE ROLE OF ARTIFICIAL INTELLIGENCE

Climate models have become indispensable to studying changes in the Earth's climate, including its future response to anthropogenic forcing. Climate modelling involves solving sets of coupled partial differential equations over the globe. Physical components of the Earth system – the atmosphere, ocean, land, cryosphere and biosphere – and the interactions between them are represented in these models and executed on high-performance supercomputers running at speeds of petaflops and beyond. Models operate by dividing the globe into grids of a specified size, defined by the model resolution. The dynamical equations are then solved to obtain output fields averaged over the size of the grid. Therefore, only physical processes operating at spatial scales larger than the grid size are explicitly resolved by the models based on partial differential equations; processes that operate at finer scales, such as clouds and deep convection, are represented by approximate empirical relationships called parameterizations. This presents at least two significant challenges:

1. While climate models have become increasingly comprehensive, grid sizes of even state-of-the-art models are no smaller than about 25 km, placing limits on their utility towards *regional* climate projections and thereby for targeted policymaking.
2. Physical processes organizing at sub-grid scales often critically shape *regional* climate. Therefore, errors in their parameterizations are known to be the source of significant uncertainties and biases in climate models. Additionally, numerous biophysical processes are not yet well understood due to the complex and non-linear nature of the interactions between the oceans, atmosphere and land.

Therefore, rapid advances are necessary to "downscale" climate model projections to higher resolutions, improving parameterizations of sub-grid scale processes and quantifying as yet poorly understood non-linear feedbacks in the climate system.

A significant bottleneck in improving model resolution is the rapid increase in the necessary computational infrastructure such as memory, processing power and storage. For perspective, an *atmosphere-only* weather model with deep convection explicitly resolved was recently run in an experimental mode with a 1 km grid size. The simulation used 960 compute nodes on SUMMIT, one of the fastest supercomputers in the world with a peak performance of nearly 150 petaflops, yet achieved a throughput of only one simulated week per day in simulating a 4 months period. A full-scale climate model, including coupled ocean, land, biosphere and cryosphere modules, must cumulatively simulate 1,000s of years to perform comprehensive climate change studies. Towards surmounting the challenge of this massive scaling up in computing power, there have recently been calls for a push towards "exascale computing" (computing at exaflop speeds) in climate research. While the technology may be within reach, practical problems abound in terms of how many centres will be able to afford the necessary hardware and the nearly gigawatt (GW) scale power requirements of exascale computing that will require dedicated power plants to enable it.

Similar bottlenecks exist for improving parameterizations of sub-grid scale processes. Satellite and ground-based measurements have produced a deluge of observational data on key climate variables over the past few decades. However, these datasets are subject to several uncertainties such as data gaps and errors arising during data acquisition, storage and transmission. The emerging challenge is to process and distil helpful information from this vast data deluge.

Towards overcoming these challenges to improving climate projections, we discuss recent advances in artificial intelligence that have enabled new insights into climate system processes. These techniques, however, are also subject to their own limitations. It is in this context that we discuss how QAI may help overcome those limitations and advance both higher resolution climate model projections and reduce their biases.

When machines learn decision-making or patterns from the data, they gain what is known as Artificial Intelligence (AI). Climate science has seen an explosion of datasets in the past three decades, particularly observational and simulation datasets. AI has seen tremendous developments in the past decade and it is anticipated that its application to climate science will help improve the accuracy of future climate projections. Recent research has shown that the combination of computer vision and time-series models effectively models the dynamics of the Earth system [3]. It is anticipated that advances in this direction would enable AI to simulate the physics of clouds and rainfall processes and reduce uncertainties in the present systems [4,5]. In addition to helping augment the representation of natural systems in climate models by using the now available high-quality data, AI has also been proposed for climate change mitigation applications [6]. Other areas where AI is playing a leading role are the technologies of carbon capture, building information systems, improved transportation systems and the efficient management of waste, to name a few [6].

There are, however, limitations to the present deep learning models; for example, their inability to differentiate between causation and correlation [7–12]. Moreover, Moore's law is expected to end by about 2025 as it bumps against fundamental physical limits such as quantum tunnelling. With the increasing demands of deep learning and other software paradigms, alternate hardware advancements are becoming necessary [13].

13.3 QUANTUM ARTIFICIAL INTELLIGENCE FOR THE SCIENCE OF CLIMATE CHANGE

AI algorithms suffer from two main problems: one is the availability of good quality data and the other is computational resources for processing big data at the scale of planet Earth. The impediments to the growth of AI-based modelling can be understood from the way language models have developed in the past decade. In the early days of their success, developments were limited to computer vision, while Natural Language Processing (NLP) lagged behind. Many researchers tried to use different algorithms for NLP problems but the only solution that broke ice was increasing the depth of the neural networks. Present-day Generative Pre-trained Transformer (GPT), Bidirectional Encoder Representations from Transformers (BERT) and Text-To-Text Transfer Transformer (T5) models are the evolved versions from that era. Maximizing gains from the rapid advances in AI algorithms requires that they be complemented by hardware developments; quantum computing is an emerging field in this regard [14,15].

Quantum Computers (QC) represent a conceptually different paradigm of information processing based on the laws of quantum physics [16]. The fundamental unit of information for a conventional/classical computer is the bit, which can exist in one of two states, usually denoted as 0 and 1. The fundamental unit of information for a quantum computer, on the other hand, is the "qubit", a two-level quantum system that can exist as a superposition of the 0 and 1 states, interpreted as being simultaneously in both states although with different probabilities. What distinguishes quantum from classical information processing is that multiple qubits can be prepared in states sharing strong "non-classical" interactions called "entanglement" that simultaneously sample a much wider informational space than the same number of bits, thereby enabling, in principle, massively parallel computation. This makes quantum computers far more efficiently scalable than their classical counterparts for certain classes of problems.

One trend of quantum computing is the race to demonstrate at least one problem that remains intractable to classical computers, but which can be practically solved by a quantum computer. Google coined this feat "quantum supremacy", and claimed, not without controversy, to achieve it with its 540 qubit Sycamore chip [17]. A research team in China introduced Jiuzhang, a new light-based special-purpose quantum computer prototype, to demonstrate quantum advantage in 2020 [18]. The University of Science and Technology of China has successfully designed a 66-qubit programmable superconducting quantum processor, named ZuChongzhi [19]. IBM plans to have a practical quantum chip containing in excess of one thousand qubits by 2023 [20].

AI on quantum computers is known as QAI and holds the promise of providing major breakthroughs in furthering the achievements of deep learning. NASA has Quantum Artificial Intelligence Laboratory (QuAIL) which aims to explore the opportunities where quantum computing and algorithms address machine learning problems arising in NASA's missions [21]. The JD AI research centre announced that they have a 15-year research plan for quantum machine learning. Baidu's open-source machine learning framework Paddle has a subproject called paddle quantum, which provides libraries for building quantum neural networks [22]. However, for practical purposes, the integration of AI and quantum computing is still in its infant stage. The use of quantum neural networks is developing at a fast pace in the research labs; however, pragmatically useful integration is in its infant stages [23,24]. The current challenges to industrial-scale QAI include how to prepare quantum datasets, how to design quantum machine learning algorithms, how to combine quantum and classical computations and identifying potential quantum advantage in learning tasks [25]. In the past five years, algorithms using quantum computing for neural networks have been developed [26,27]. Just as the open-source TensorFlow, PyTorch and other deep learning libraries stimulated the use of deep learning for various applications, we may anticipate that software, such as TensorFlowQ (TFQ), QuantumFlow and others, already in development will stimulate advances in QAI.

13.3.1 COMPLEX PROBLEMS IN EARTH SYSTEM SCIENCE: POTENTIAL FOR QAI

QAI can be used to learn intelligent models of earth system science bringing new insights into the science of climate change. It can play an essential role in designing climate change strategies based on improved, high-resolution scientific knowledge powered by QAI. Recent studies (for example, Ref. [28]) have attempted to develop physics schemes based on deep learning. However, these are largely proof-of-principle studies in nascent stages. Challenges such as the spherical nature of the data over Earth, complex and non-linear spatio-temporal dynamics and others exist in AI for improved climate models. Various techniques such as cubed spheres and tangent planes have been proposed to address the spatial errors arising out of sphericity. QAI can further develop advanced physical schemes using AI by incorporating high-resolution datasets, more extended training, and hyperparameter optimization. A necessary condition for the quantum speedup of classical AI is that the task in question can be parallelized for training. Present libraries such as TensorFlow and PyTorch offer both data and model parallelism capabilities. They have also been released for quantum computers and need to be further developed for industrial-scale quantum computers of the future.

13.3.2 TECHNOLOGICAL SOLUTIONS FOR IMPLEMENTING QAI

Figure 13.1 shows the technological solutions for implementing QAI in climate science. The main technologies to empower QAI for climate science include the resources in the two GitHub repositories on awesome Quantum Machine Learning (QML) [29,30], QML [31], Pennylane [32], Quantum Enhanced Machine Learning (QEML) [33], TensorFlowQ [34] and NetKet [35].

13.3.3 CASE STUDY ON THE USE OF QAI FOR CLIMATE SCIENCE

We demonstrate an example of the application of QAI for land-use land-cover classification on the UC Merced dataset. The dataset is first transformed to

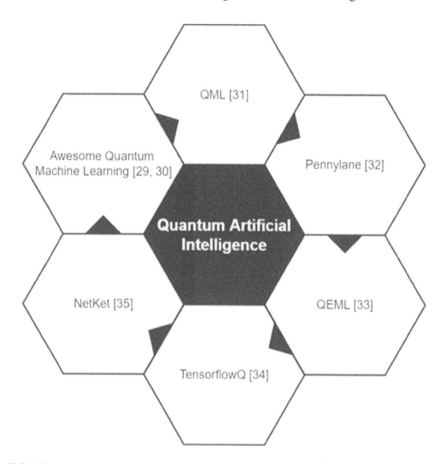

FIGURE 13.1   Technologies on QAI for the science of climate change

quantum data using Pennylane library and the training is performed. The code can be found on the GitHub repository for this article at https://github.com/manmeet3591/qai_science_of_climate_change.

Figure 13.2 shows the working of the case study. Initially, satellite data is transformed as quantum data using Pennylane. Then, we apply the deep convolutional classification algorithm for land-use land-cover classification. The output classes consist of forests, agricultural fields, etc.

13.4   CONCLUSIONS AND FUTURE WORK

Simulation studies are used to understand the science of climate change and are computationally expensive tools to understand the role of various forcings on the climate system. For example, recently, a study in the journal Science Advances showed how volcanic eruptions could force the coupling of El Nino Southern Oscillation and the South Asian Monsoon systems. Works of such kind are critical in advancing the understanding of the climate system and its response to various forcings. However,

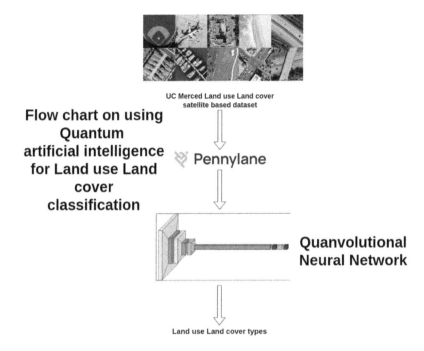

FIGURE 13.2   A case study on using QAI for land-use land-cover classification

they are computationally demanding and are time-consuming to complete. Large ensemble climate simulations is an area that requires further work using quantum computing and QAI. QML can play an important role, especially in pattern recognition for weather and climate science, with problems that will benefit the most from quantum speedup being those that are inherently parallelizable.

However, various challenges present themselves in designing and operating useful quantum computers. State-of-the-art implementations of quantum computers today can control and manipulate on the order of 100 qubits, whereas it is estimated that any real-world applications where quantum computers can reliably outperform classical computers would require on the order of a million qubits. This presents a formidable technological challenge. Additionally, entanglement – the heart of quantum computing – is a fragile resource prone to being destroyed with even the slightest disturbance (called "decoherence"). Therefore, operational quantum computers may be several years into the future. Yet, given their potential to affect a genuine paradigm shift, much effort has been invested in exploring their application to various fields and the focus of the present article concerns their possible climate science applications.

In short, QAI is projected to be a powerful technology of the future. Developments in the field of QAI include the development of, both, computer vision and sequence algorithms capable of being implemented on large quantum computers. All these advancements would be driven by one factor, i.e., the development of high-performance quantum computing hardware.

13.4.1 Software Availability

We have released the code for demonstrating the application of QAI on land-use land-cover classification dataset from the UC Merced dataset at: https://github.com/manmeet3591/qai_science_of_climate_change

REFERENCES

[1] Wichert, A., 2020. *Principles of Quantum Artificial Intelligence: Quantum Problem Solving and Machine Learning.* World Scientific, Singapore.
[2] O'Brien, K.L., 2016. Climate change and social transformations: is it time for a quantum leap? *Wiley Interdisciplinary Reviews: Climate Change*, 7(5), pp. 618–626.
[3] Weyn, J.A., Durran, D.R. and Caruana, R., 2020. Improving data-driven global weather prediction using deep convolutional neural networks on a cubed sphere. *Journal of Advances in Modeling Earth Systems*, 12(9), p. e2020MS002109.
[4] Singh, M., Tuli, S., Butcher, R.J., Kaur, R. and Gill, S.S. 2021. Dynamic shift from cloud computing to industry 4.0: Eco-friendly choice or climate change threat, In P. Krause and F. Xhafa (eds.), *IoT-Based Intelligent Modelling for Environmental and Ecological Engineering, Lecture Notes on Data Engineering and Communications Technologies*, vol. 67, Springer, Cham.
[5] Jiang, W., Xiong, J. and Shi, Y., 2021. A co-design framework of neural networks and quantum circuits towards quantum advantage. *Nature Communications*, 12(1), 1–13.
[6] Rolnick, D., Donti, P.L., Kaack, L.H., Kochanski, K., Lacoste, A., Sankaran, K., Ross, A.S., Milojevic-Dupont, N., Jaques, N., Waldman-Brown, A. and Luccioni, A., 2019. Tackling climate change with machine learning. arXiv preprint arXiv:1906.05433.
[7] Reichstein, M., Camps-Valls, G., Stevens, B., Jung, M., Denzler, J. and Carvalhais, N., 2019. Deep learning and process understanding for data-driven earth system science. *Nature*, 566(7743), pp. 195–204.
[8] Rasheed, A., San, O. and Kvamsdal, T., 2020. Digital twin: Values, challenges and enablers from a modeling perspective. *IEEE Access*, 8, pp. 21980–22012.
[9] https://twitter.com/michaelemann/status/1304039845598199809?lang=en
[10] Schiermeier, Q., 2010. The real holes in climate science. *Nature News*, 463(7279), pp. 284–287.
[11] Britt, K.A. and Humble, T.S., 2017. High-performance computing with quantum processing units. *ACM Journal on Emerging Technologies in Computing Systems (JETC)*, 13(3), pp. 1–13.
[12] Singh, M., Krishnan, R., Goswami, B., Choudhury, A.D., Swapna, P., Vellore, R., Prajeesh, A.G., Sandeep, N., Venkataraman, C., Donner, R.V. and Marwan, N., 2020. Fingerprint of volcanic forcing on the ENSO–Indian monsoon coupling. *Science Advances*, 6(38), p. eaba8164.
[13] Marcus, G., 2018. Deep learning: A critical appraisal. arXiv preprint arXiv:1801.00631.
[14] Eeckhout, L., 2017. Is Moore's law slowing down? What's next? *IEEE Micro*, 37(4), pp. 4–5.
[15] Singh, M., Kumar, B., Niyogi, D., Rao, S., Gill, S.S., Chattopadhyay, R. and Nanjundiah, R.S., 2021. Deep learning for improved global precipitation in numerical weather prediction systems. arXiv preprint arXiv:2106.12045.
[16] Gill, S.S., Kumar, A., Singh, H., Singh, M., Kaur, K., Usman, M. and Buyya, R. 2022, Quantum computing: A taxonomy, systematic review and future directions. *Softw Pract Exp*, 52(1), pp. 66–114.
[17] Arute, F., Arya, K., Babbush, R. et al., 2019. Quantum supremacy using a Programmable superconducting processor. *Nature*, 574, pp. 505–510. https://doi.org/10.1038/s41586-019-1666-5

[18] Zhong, H.S., Wang, H., Deng, Y.H. et al., 2020. Quantum computational advantage using photons. *Science*, 370(6523), pp. 1460–1463 https://doi.org/10.1126/science.abe8770

[19] Wu, Y., Bao, W.S., Cao, S. et al., 2020. Strong quantum computational advantage using a superconducting quantum processor. *Physical Review Letters*, 127(18), p. 180501. https://doi.org/10.1103/PhysRevLett.127.180501

[20] IBM's roadmap for scaling quantum technology, https://www.ibm.com/blogs/research/2020/09/ibm-quantum-roadmap/

[21] NASA Quantum Artificial Intelligence Laboratory. https://ti.arc.nasa.gov/tech/dash/groups/quail/

[22] Paddle Quantum. https://github.com/PaddlePaddle/Quantum

[23] Beer, K., Bondarenko, D., Farrelly, T. et al., 2020. Training deep quantum neural networks. *Nature Communication*, 11, p. 808. https://doi.org/10.1038/s41467-020-14454-2

[24] Kong, F., Liu, X.Y., and Henao, R., 2021. Quantum tensor network in machine learning: An application to tiny object classification. arXiv preprint arXiv:2101.03154.

[25] Huang, HY., Broughton, M., Mohseni, M. et al., 2021. Power of data in quantum machine learning. *Nature Communications*, 12, p. 2631 (2021). https://doi.org/10.1038/s41467-021-22539-9

[26] Huntingford, C., Jeffers, E.S., Bonsall, M.B., Christensen, H.M., Lees, T. and Yang, H., 2019. Machine learning and artificial intelligence to aid climate change research and preparedness. *Environmental Research Letters*, 14(12), p. 124007.

[27] Jiang, W., Xiong, J. and Shi, Y., 2021, When machine learning meets quantum computers: A case study. In *2021 26th Asia and South Pacific Design Automation Conference (ASP-DAC)* (pp. 593–598). IEEE.

[28] Rasp, S., Pritchard, M.S. and Gentine, P., 2018. Deep learning to represent subgrid processes in climate models. Proceedings of the National Academy of Sciences, 115(39), pp. 9684–9689.

[29] https://github.com/krishnakumarsekar/awesome-quantum-machine-learning

[30] https://github.com/artix41/awesome-quantum-ml

[31] https://www.qmlcode.org/

[32] Bergholm, V., Izaac, J., Schuld, M., Gogolin, C., Alam, M.S., Ahmed, S., Arrazola, J.M., Blank, C., Delgado, A., Jahangiri, S. and McKiernan, K., 2018. Pennylane: Automatic differentiation of hybrid quantum-classical computations. arXiv preprint arXiv:1811.04968.

[33] Sharma, S., 2020. Qeml (quantum enhanced machine learning): Using quantum computing to enhance ml classifiers and feature spaces. arXiv preprint arXiv:2002.10453.

[34] Broughton, M., Verdon, G., McCourt, T., Martinez, A.J., Yoo, J.H., Isakov, S.V., Massey, P., Halavati, R., Niu, M.Y., Zlokapa, A. and Peters, E., 2020. Tensorflow quantum: A software framework for quantum machine learning. arXiv preprint arXiv:2003.02989.

[35] Carleo, G., Choo, K., Hofmann, D., Smith, J.E., Westerhout, T., Alet, F., Davis, E.J., Efthymiou, S., Glasser, I., Lin, S.H. and Mauri, M., 2019. NetKet: A machine learning toolkit for many-body quantum systems. *SoftwareX*, 10, p. 100311.

14 Quantum Computing-Based Optimization in Depression Detection Using Speech

Surbhi Sharma
Jawaharlal Nehru University

CONTENTS

14.1 INTRODUCTION

Feature selection has always been the most important concern in solving machine learning problems. Dimension reduction is a major concern when data is having multiple features and the sample size is small. There are two types of feature selection techniques available – *filter* and *wrapper*. In filter feature selection, the relevance of a feature is considered with respect to a class label based on statistical descriptors. However, in wrapper feature selection, a relevant subset of feature is taken based on the evaluation criteria of the learning algorithm. But these algorithms have not been proved fruitful in all machine learning problems. In random feature selection, the major drawback is that those features that are selected/removed cannot be selected/removed in subsequent iterations. Therefore, there is a need to search for such a feature selection algorithm

DOI: 10.1201/9781003250357-14

that is exploratory in nature rather than sequential in nature as well as it should be efficient in time and space complexity [1]. In the lineage of exploratory algorithms, Whale optimization algorithm [2], Particle swarm optimization [3], and Genetic algorithm [4] are evident. But classic evolutionary algorithms are computationally intensive in time and space. If the nature-inspired algorithms [5] are combined with quantum computing, those algorithms have been found effective in feature selection as compared to the previous state of the art. The main objective of the work is to discuss the various types of nature-inspired algorithms and to depict how will the collaboration of quantum computing with the evolutionary algorithm help in obtaining a relevant subset of features. It is to accentuate that quantum whale optimization can play a vital role in feature selection of the Depression Dataset. To illustrate this, the traditional whale optimization algorithm has been combined with quantum computing to explore the feature space and thereby reduce the high dimensional feature space to an optimal one in less time complexity. Initially, the various nature-inspired algorithms like Genetic algorithm, Particle swarm optimization, Ant colony optimization, and traditional Whale optimization algorithm have been discussed. These evolutionary algorithms have been enumerated to highlight that if evolutionary algorithms are combined with quantum computing, these can play a vital role in feature selection. Then, quantum whale optimization and its role in feature reduction have been explained.

14.2 GENETIC ALGORITHM

The input to the Genetic Algorithm (GA) [4] is a collection of chromosomes that represent each individual. Each individual is evaluated on the basis of the fitness criteria. Each chromosome is the representation of an individual. The initial chromosomes that are fed to the GA [4] are the potential solution and are randomly generated. The GA then evaluates each chromosome and evaluates it on the basis of a fitness function. The candidates having optimal fitness value are passed on to the next generation. The promising candidates are passed on to the next generation by reproducing their copies, but copies are not a perfect representation of their parents, copies are with random changes. Then, those offspring become potential candidate for optimal solution that is passed to the next generation that is again evaluated on the basis of the fitness function. The nature of the fitness function is decided according to the problem statement. The candidate chromosomes which are giving poor results are removed in subsequent generations that are based on random selection. Selection criteria for passing candidates to the next generation are given as below: [4]

- More fit individuals are likely to be selected.
- **Roulette Wheel Selection:** The candidates based on the deviation from the fitness value of the competitor are passed on to the next generation.
- We can scale the fitness function based on the fact that many individuals are having fitness value in close range.

Randomness in the search space is provided by crossover and mutation operators. These operators imply that in the next generation, new candidate solutions are provided on the basis of the value of the fitness function.

- **Crossover:** Two parents are taken who are good performing candidates. Swapping one part of the chromosome of one parent with another part of the parent is allowed. Two offspring that are produced are passed to the next generation. The selection of the crossover operation is based on some probability for each chromosome.
- **Mutation:** The chromosome is chosen randomly and mutation is done at one point.

The disadvantage of the GA is that certain problems cannot be solved by the GA. It takes computationally intensive time to reach a global optimal. It does not always ensure an optimal solution. Therefore, it may not be always a good choice for the feature selection.

14.3 PARTICLE SWARM OPTIMIZATION

The particle updates its position from one iteration to another iteration. Each particle tries to move in direction of its own best solution as well as the global best solution as represented by Eq. (14.1).

$$\text{pbest}(i,t) = \arg\min[Y(f_i(k)], i \in \{1,2,....,N_f\}$$

$$k = 1,....,t$$

$$\text{gbest}(t) = \arg\min[Y(f_i(k))] \tag{14.1}$$

$$i = 1,....,N_f$$

$$k = 1,.....,t$$

where pbest refers to the individual best of the particle; however, gbest refers to the global best among particles. In the above-mentioned equation, Y refers to fitness function, t is iteration number, and i stands for the particle number. Nf means the total number of particles. The f variable defines the current position of the particle. $r1$ and $r2$ here are uniformly random variables over 0 and 1 and $c1$ and $c2$ are called acceleration coefficients. These are constant parameters. The updated velocity [3] and the updated position [3] of the particle are given by following Eq. (14.2) as follows:

$$V_i(t+1) = \omega V_i(t) + c_1 r_1 \left(\text{pbest}(i,t) - P_i(t)\right) + c_2 r_2 \left(\text{gbest}(t) - P_i(t)\right)$$
$$P_i(t+1) = P_i(t) + V_i(t+1) \tag{14.2}$$

where ω defines the inertia component of the velocity which is responsible for maintaining the balance between exploration and exploitation. The second factor is responsible for the cognitive move which is responsible for the individual move to imitate its best position, and the last factor in the equation qualifies for the collaborative effort of the particles to move in direction of the best move.

In Ref. [6], the Particle Swarm Optimization Algorithm (PSOA) is explored. The important steps of this algorithm are given as follows. In this algorithm, for each particle, initialize the particle's position and mark their pbest to their initial position and gbest to the minimum value of all swarm. For each particle, pick two random numbers between 0 and 1. Update particle's velocity and position. Update the best-known position of particles as well as update the swarm's best-known position. Repeat this until the termination condition is met. In the end, output gbest as the required solution.

Step 1: Initialization
For each particle $i = 1,...,N_T$, do

a. Initialize the particle's position with a uniform distribution as $T_i(0) \sim U(\text{LB}, \text{UB})$, where LB and UB represent the lower and upper bounds of the search space, respectively.
b. Initialize pbest to its initial position: pbest $(i,0) = T_i(0)$.
c. Initialize gbest to the minimal value of the swarm: gbest $(0) = \text{argmin} f[P_i(0)]$.
d. Initialize velocity: $V_i \sim U(-|\text{UB} - \text{LB}|, |\text{UB} - \text{LB}|)$.

Step 2: Repeat until a termination criteria is met.
For each particle $i = 1,...,N_T$, do

a. Pick random numbers: $r_1, r_2 \sim U(0,1)$.
b. Update particle's velocity. See Eq. (14.2).
c. Update particle's position. See Eq. (14.2).
d. If $f[T_i(t)] < f[\text{pbest}(i,t)]$, do
   i.   Update the best-known position of particle i: pbest $(i,t) = T_i(t)$.
   ii. If $f[T_f(t)] < f[\text{gbest}(t)]$, update the swarm's best-known position: gbest $(t) = T(t)$.
e. $t \leftarrow (t+1)$;

Step 3: Output gbest (t) that holds the best-found solution.
In Ref. [3], PSOA steps are presented. These steps and their sequences are drawn in a flowchart as shown in Figure 14.1. First, the initialization of position and velocity is done randomly, local best to its initial local best and global best to the local best are done. Then, update particle velocity and update particle position. Correspondingly, update local best and global best. When the stopping criteria is met, global best would be output as the solution.

14.4   ANT COLONY OPTIMIZATION

Ant Colony Optimization (ACO) [7] is a metaherustic algorithm that can be used for all types of optimization problems. First of all, ants are in their corresponding nest. Now they start their search in direction of the food. Earlier, there was no pheromone. Pheromone is a chemical that ants leave while trailing the path so that followers ants can follow the path. The pheromone actually represents the strength of the path that decides the probability of the selection of the path to choose from among many paths

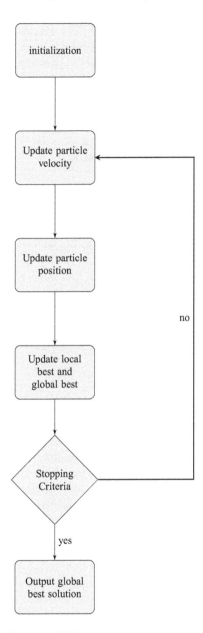

FIGURE 14.1   The flow diagram of PSOA

available. Ants choose their path with some probability. Longer paths take longer time to reach food compared to shorter paths. While returning back to their home, as the evaporation rate is smaller of the pheromone compared to the evaporation rate of the pheromone of the longer path, therefore, still the probability of selection of the shorter path remains higher compared to the longer path. The pheromone concentration increases more on the shorter path while ants are returning over the shorter path.

Therefore, a pheromone updation is required. The concentration of the pheromone gets reduced in the long path mitigating its probability of selection. That leads the ants to follow that path only. We can transform this problem in form of a graph where the source home is taken as a source vertex and where the food is lying is treated as the destination vertex. Now, there might be multiple paths between the source vertex and the destination vertex. Suppose there are two edges e1 and e2 between the source and destination vertex having their respective strengths known to be as pheromone. The strength will then decide the probability of selection of the path. There are three stages involved in this process as described below:

$$Pi = \frac{Ri}{R1 + R2}; i = 1, 2 \tag{14.3}$$

where $R1 > R2$ decides the probability of R1 is greater than R2; otherwise, the probability of R2 is greater than R1 and vice-versa. Equation 14.3 represents probability for the selection of each path between source vertex and destination vertex.

$$Ri \leftarrow Ri + \frac{K}{Li} \tag{14.4}$$

Here, the updation is done in accordance with the path length of the edge. The pheromone value for each edge is updated as above, where K serves as a model parameter.

$$Ri \leftarrow (1 - v) * Ri \tag{14.5}$$

When the ants return over the path after carrying food, the updation of the pheromone is carried out according to the above-mentioned Eq. (14.5). Here, the value of v varies from 0 to 1. At every iteration, the ants are kept at the source vertex, then they move from the source vertex to the destination vertex. The updation of the pheromone is carried out on the basis of Eq. (14.4) after reaching the destination. While returning from the path, they choose their return path on the basis of the updation of Eq. (14.5) which is based on the rate of evaporation of the pheromone.

The algorithm for ACO [7] is described as follows (Figure 14.2):

In Figure 14.3, the steps of ACO are presented. These steps and their sequences are drawn in a flowchart, as shown in Figure 14.3. First, ACO parameters are done. Construct the solution using probability distribution. Local updation of pheromone

---

Algorithm 1: Ant Colony Optimization

Initialize necessary parameters and pheromone trials;
while *not termination* **do**
  Generate ant population;
  Calculate fitness values associated with each ant;
  Find best solution through selection methods;
  Update pheromone trial
end

FIGURE 14.2 The algorithm of ACO [7]

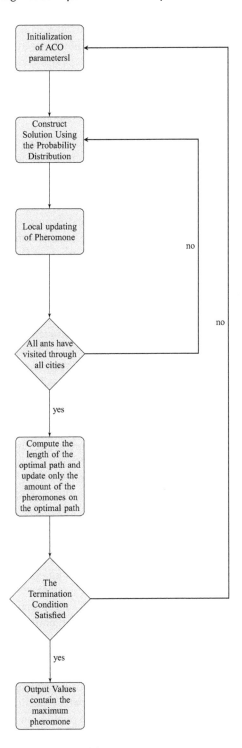

FIGURE 14.3   The flow diagram of ACO

is done. When all ants have visited through all the cities, compute the length of the optimal path and update only the amount of the pheromones on the optimal path. When the termination condition is met, output the values containing the maximum pheromones.

14.5 WHALE OPTIMIZATION ALGORITHM

Whale Optimization Algorithm [2] explores the search space in order to find an optimal solution. It first tries to explore the search space and then tries to exploit it. Exploitation means targeting towards the prey in more proximity of it. It has been found that evolutionary algorithm has performed fairly better as compared to its respective counterparts like GA, PSOA, and ACO. As Curse of Dimensionality is a major concern where the sample size is small and features are multidimensional. There is a need to seek for an appropriate feature selection method in that case. When we talk about making a good decision model for Depression Detection, where people are trying to adopt a multimodality approach to enhance the performance of the decision model, it becomes essential to explore an exhaustive set of features for the unimodal approach only as multimodal approach is computationally intensive. People are trying to explore an exhaustive set of features from speech behaviour, speech prosody, eye behaviour, and head behaviour. As data size is small and features are extensive in nature, there is an essential need to select a handful of features which would suffice in discriminating depressed patients from control ones so that time and space complexity of a decision model can be reduced. As said earlier, there are many feature selection techniques available which are having their own limitations. Some filter feature selection techniques do not consider correlation among features while selecting them. However, others only consider the correlation with class label. Therefore, there is a need to look for such a feature selection technique which can eliminate the limitations of various feature selection techniques. We should look for such an evolutionary nature-inspired optimization algorithm. The whale optimization algorithm considers n whales of d dimension that is equivalent to the feature dimension of the data. Then, n whales are represented in the binary representation. The binary value 1 which is considered to be the corresponding feature value is selected; on the other hand, binary value 0 which is considered to be that feature value is not selected. The n whales are evaluated on the basis of the fitness criteria of the decision model. The whale whose fitness value is an optimal one that is categorized as the best whale population. The process is repeated for the maximum number of generations. In each generation, either of the three steps is performed before evaluating the n whales population. The first one is exploration, the second one is the shrinking phase, and the last one is the spiralling phase. In the exploration phase, a random whale is chosen and it is crossed over and mutated over the current whale position. However, in the shrinking phase, the best whale population is crossed and mutated over the current whale position. In the spiralling phase, the distance vector which calculates the difference between the best whale position and current whale position is calculated and the position of the whale population is updated as follows.

$$D = W^* - W^i \qquad (14.6)$$

The updated position of the population for spiralling is given as follows:

$$W^i = \mathbf{D} \cdot e^{pl} \cdot \cos(2\pi l) + W^*$$ (14.7)

where p is having a random value between 0 and 1 that decides in each generation whether exploration, shrinking and spiralling would be decided for updation, as well as it is decided by the value of a coefficient A which is given as below. In other words, p and A are helping in creating randomness in our search space so that we can reach our optimal solution space. The value of l belongs from a range of −1 to 1.

$$A = 2q \cdot r - q$$ (14.8)

where q is given as below whose range varies from 0 to 2.

$$q = 2 - m * 2 / \text{maxGen}$$ (14.9)

where m refers to a current generation and maxGen refers to the total number of generations. The above-mentioned whale optimization algorithm takes more time to explore and converge to an optimal solution. Therefore, there is a need to investigate an amalgamation of quantum computing with the whale optimization algorithm which has been found to effectively converge to an optimal solution in less time and space complexity rather than the whale optimization algorithm.

14.6 QUANTUM COMPUTING-BASED WHALE OPTIMIZATION ALGORITHM FOR FEATURE SELECTION

Quantum computing defines the state of the computer as Q-bit rather than bit representation. Q-bit refers to quantum-bit. A quantum bit is defined as the superposition of two states. The state of a quantum computer is defined as the probability of occurrence in being one of the states either 0 or 1 state.

$$|\Psi\rangle = a|0\rangle + b \mid 1\rangle$$ (14.10)

where

$$|a|^2 + |b|^2 = 1$$ (14.11)

Each individual modulus value defines the probabilistic approach in being one of the states. Suppose if our vector is of length d. Then, we can conclude that the vector can be in one of the 2^d states. The entire d length vector can be transformed into a binary length vector. On the basis of a specific threshold value, we can decide whether the particular feature will be selected or not. The threshold value is taken as random.

if $|a_i|^2$ <*threshold* **then**

$$y_i \leftarrow 1$$

else

$$y_i \leftarrow 0$$

end if

To update each population in every generation mutation, selection and crossover are performed. The updation can be done in many ways. The mutation can be done by interchanging the either bit position of two populations. However, the crossover can be performed by swapping the bit position either random or at n point position between two parents to give rise to the two children. There is a need to balance the exploitation and exploration in an evolutionary algorithm that too in a shorter span of time. For this, the quantum gate operators are applied at the probability amplitude of each amplitude. There are various types of quantum gates that can be used. The one we used while performing experiments is the rotation gate operator.

$$U(\varphi) = \begin{bmatrix} \cos(\varphi) & -\sin(\varphi) \\ \sin(\varphi) & \cos(\varphi) \end{bmatrix} \tag{14.12}$$

where φ represents the angle of rotation and the direction of the rotation that is decided from the lookup table [8].

14.6.1 Quantum Computing Whale Optimization Algorithm Flow

First, we take n whale population, where n is a random number chosen by us. Each population is taken to be of d dimension vector. The d dimension is represented in the form of Q-bits. For the exploration phase, we pick a random population that will be supposed as the best individual population initially in direction of which the search for optimal solution proceeds.

- **Crossover:** In each generation, the crossover is performed between mutated random population and the current individual.
- **Shrinking:** In each generation, the crossover is performed between the best individual selected in accordance with fitness value and the mutated form of the current individual.
- **Spiral Phase:** Spiral phase is decided by the Eqs. (14.1) and (14.2) mentioned above.

In each generation, the quantum gate operator is applied to balance out the exploration and exploitation. The fitness value that we have considered is the superposition of error and minimum number of features (Figure 14.4).

First, we have to initialize the whale population of size n where d is the dimension of each whale population. Each whale dimension is represented in the form of quantum bits. Each whale population is collapsed to binary population. Calculate the fitness value of each binary vector. Store the best fitness vector and the index of the best fitness vector. For each maximum generation of 100, for each individual, choose a random value of p and update A. On the basis of the values of p and A, it would be decided whether that particular population would undergo spiral, shrinking, and

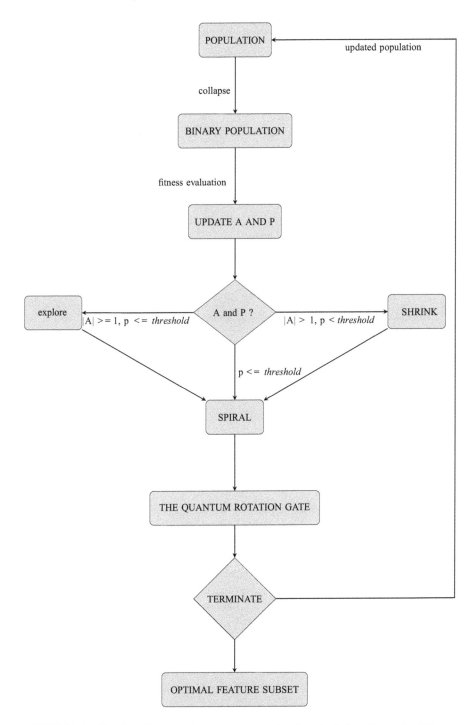

FIGURE 14.4   The flow diagram of quantum computing-based whale optimization [1]

Algorithm 1 Quantum Whale Optimization Algorithm

$t \Leftarrow 1$

$T \Leftarrow maxgen$

Initialize a feature population of size n

$$M(t) = \left\{ M^1(t), M^2(t), \ldots, M^i(t), \ldots, M^n(t) \right\}^T$$

where $M^i(t) = \left\{ M^{i1}(t), M^i(t), \ldots, M^{ij}(t), \ldots, M^{id}(t) \right\}$, d : $dim.$ of quantum

individual $M^{ij} = \begin{bmatrix} a_{ij} \\ b_{ij} \end{bmatrix}$ is a Q - bit, s.t. $|a_{ij}|^2 + |b_{ij}|^2 = 1$

Observe/Collapse M(t) to give the binary population B(t)

Calculate Fitness value for each binary feature vector $B^i(t), i = 1, 2, 3 \ldots n$

$B^*(t) \leftarrow$ BestFitnessVector(B(t))

bestIndex = BestFitnessIndex (B(t))

$M^*(t) \leftarrow M^{bestindex}(t)$

While $t < T$

For each individual, $i = 1, 2, 3, \ldots n$

Choose a random value p and update A

If p < threshold

If $abs(A) \geq 1$ //Explore

$M^T(t) \leftarrow$ Tournament Selection (M(t))

$M^T(t) \leftarrow$ Mutation $(M^T(t))$

$M^i(t) \leftarrow$ Crossover(Mutation $(M^i(t))$, $M^t(t))$

If $abs(A) < 1$ //Shrink

$M^i(t) \leftarrow$ Crossover(Mutation($M^*(t))$, $M^*(t))$

If $p \geq$ threshold //Spiral

$D \leftarrow distance (M^i(t), M^*(t))$

$M^i(t) \leftarrow$ Update i^{th} individual

End For

Update M(t) using rotation gate

Observe/Collapse M(t) to give the binary population B(t)

$B^*(t) \leftarrow$ BestFitnessVector (B(t))

bestIndex = BestFitnessIndex (B(t))

$W^*(t) \leftarrow W^{bestindex}(t)$

End While

Return optimal Feature Vector $B^*(t-1)$

FIGURE 14.5   The algorithm of quantum computing-based whale optimization [1]

exploration. Update each whale population using rotation gate operator. Observe each whale population to give binary population. Calculate the fitness value of each population. Store the best fitness vector and index of the same. Now, repeat the entire procedure of spiralling, shrinking, and exploration (Figure 14.5).

As we can see QWOA has performed better than state of the art as well as in respect of the number of features. The classifier used in evaluating the performance using QWOA is K-nearest neighbour. Similar work can be tried on the depression dataset as it is multidimensional and there is also an essential need to reduce the dimensionality of the features so that the decision model can be strengthened in terms of time complexity.

14.7   DEPRESSION

In recent times, depression has emerged as the most challenging health concern for our society. Physicians define depression as a psychiatric mood disorder in which an individual is unable to cope with stressful life events and is troubled with persistent feelings (for longer than at least two-week period) of sadness, negativity, and difficulty in carrying out everyday responsibilities. Depression has been identified as the fourth biggest cause of disability by World Health Organization (WHO) in 2012 [9]. Moreover, it is predicted to be the second leading cause of disability by 2030. In the year 2014, WHO declared that 800,000 people die from suicides every year. Out of the total global depressed population, 18% belongs to India [10]. As per a report, 4.5% of India's population has been diagnosed with depression in the year 2015. The Associated Chambers of Commerce and Industry of India reports [10] that 42.5% of corporate employees in India suffer from depression. It has been reported that depression and hectic work hours drive at least one Delhi cop to suicide every month [11]. The economic impact of depression is affected by the non-productivity of individuals and increasing suicide rates. As depressed people are more prone to suicidal tendency, therefore, it becomes essential to diagnose depression at an early stage so that timely prevention can be enforced and well-wishers of depressed people can be informed. The American Psychiatric Association publishes a well-acclaimed manual to diagnose mental disorders – The Diagnostic and Statistical Manual of Mental Disorders (DSM) [12]. The DSM gives standard symptoms to identify depression, viz. psycho-motor retardation or agitation, fatigue or loss of energy, insomnia or hypersomnia, significant weight loss or weight gain, recurrent thoughts of death, and feeling of worthlessness. There are some well-known subjective measures to diagnose depression used by clinicians. One of the measures is the Hamilton Rating Scale for Depression (HAMD) [13] which is an interview-style measure used by clinicians. HAMD consists of 21 questions. Each question has a score based on the importance of symptom level. Based on the responses, an average score is computed and depression is categorized as normal, mild, moderate, severe, and very severe. There are some self-assessment measures as well. The Beck Depression Index (BDI) [14] and the eight-item Patient Health Questionnaire depression scale (PHQ-8) [15] are self-assessment measures based on self likes and dislikes. PHQ-8 consists of eight questions. A self-assessment measure takes less time to perceive depression in comparison to a clinical-based questionnaire.

Depression is a mood disorder that mostly goes undetected and untreated. There is an urgent need to build an objective system that can detect depression in time, to avoid the direct consequences of suicide. In literature, speech, speech content, images, and videos are some of the modalities that have been explored independently (unimodal) as well as in combination (multimodal), to detect depression. As multimodal systems are not cost-effective and are computationally intensive, there is a need to strengthen the unimodal systems. Speech is a non-invasive medium which is strongly correlated with the depressed state of mind and is simple and cost-effective. Though some feature extraction methods have been explored in speech-based systems, an exhaustive set of possible temporal and spectral features has not been studied. We have investigated an exhaustive set of temporal and spectral features, along with a hybrid of the two for speech-based depression detection. It is also desirable to determine a smaller set of relevant features to overcome the problem associated with the small sample size problem.

14.7.1 Objective Measures for Depression Detection

Subjective measures are biased as they are dependent upon the clinician's inter-pretation of the patient's response. Largely patients are not cooperative and do not express themselves honestly with the clinician; scores are also unreliable as at times the patients are familiar with the repeated questions. So, there is a need to explore some objective biomarkers which are reliable and which can quantify the extent of the problem that an individual is facing. In recent years, researchers have explored mental state in conjunction with speech [16], gestures [17], gait [18], video [19], and body language [17] and have also used Natural Language Processing [20] techniques and found a correlation between the depression severity and the result of these measures. The depression detection systems that combine the features of two or more modalities (multimodal systems) are capable of capturing the psycho-motor as well as the psycho-cognitive impairments that are manifested in people suffer-ing from depression. Such systems are able to capture the various aspects of behav-ioural markers of depression, and they take the advantage of the individual modality. But, capturing multimodal data is tedious and expensive due to the involvement of the visual-recording equipment and the reluctance of the individual in the pro-cess where his/her identity is exposed. Further, these methods are computationally intensive; therefore, there is a need to strengthen the unimodal systems, which are cost-effective and are computationally inexpensive. Among the various unimodal systems explored by researchers, speech-based depression detection involves low cost of acquiring the data and low cost of the learning model. Also, this system has the added advantage of being non-invasive and does not expose the individual's identity. Therefore, a unimodal system based on speech is a strong contender for depression detection. In literature, many features are extracted from speech and their correlation has been established with depression detection. Speech is a non-invasive medium and is strongly correlated with depression symptoms. It is well understood that mental stress has an immediate impact on the tenseness of the vocal chords and the tract [21], resulting in a change in the quality of speech of an individual. The vocal tract and chords get tensed due to cognitive impairments and stress exposure associated with depression [21], leading to changes in acoustics, articulation, and rate of speech. Therefore, speech can be used as an objective biomarker to diagnose depression. Speech is a medium to understand the state of the mind of a person and this opens up possibilities for exploring various features that can reliably predict the severity of depression of an individual. Speech has shown positive results as a discriminative measure which can distinguish a depressed individual from the con-trol [16, 19, 21–24]. Various feature extraction methods [25] are employed to extract features like Mel Frequency Cepstral Coefficients (MFCCs), formants, Normalized Amplitude Quotient (NAQ), Quasi-open Quotient (QOQ), and fundamental fre-quency (F0). Feature extraction is the process of transforming raw speech signal into more compact representation with limited redundancy. Features from speech signal can be extracted in Time domain (Temporal features), Frequency domain (Spectral features), and Spectro-temporal domain. Some of the time domain features suggested in literature [26] to identify a depression state are: short time energy [22], short time Zero Crossing Count (ZCC)1 [16], pitch period [22], Linear Predictive Coefficients

(LPC) [27], and loudness [22]. Some of the spectral features (frequency domain features) [27] suggested in literature are MFCCs [26], Spectral-centroid [27], Empirical Mode Decomposition (EMD) [28], Energy slope [27], Jitter [29,30], Shimmer [29,30] Harmonic to Signal Noise Ratio (HNR) [31] and Power Spectral Density (PSD) [8]. Some of the spectro-temporal features suggested in literature are Maxima Dispersion Quotient (MDQ) [32], and Peakslope [33] which measure both temporal as well as spectral information. Though researchers have explored a few temporal/spectral features, no work has been done that explores an exhaustive set of features. We have explored the exhaustive set of temporal, spectral and spectro-temporal features. We have conducted experiments to understand the importance of just the temporal or the spectral features, and also a combination of all features, in order to establish the role of these features in depression detection. Speech-based unimodal depression detection system, which is non-invasive, involves low cost and computation time in comparison to multimodal systems. The performance of any decision system mainly depends on the choice of feature selection method and the classifier.

14.7.2 RELATED WORK

Speech is a natural, non-invasive evidence that has been investigated for depression over the years and has been found indicative in classifying depressed people from the control group. The tenseness of the vocal tract and the vocal cords in depressed people translates to a set of features that are well-known and have been investigated over the years. France et al. [24] worked with the fundamental frequency (F0), Amplitude Modulation (AM), formants, and Power Spectral Density (PSD) for discriminating control, dysthymic, and major depressed persons. Formant frequencies and PSD were successful with 0.94 accuracy to distinguish between the control and depressed female patients and 0.82 between the male patients. Cummins et al. [16] have shown merits of voiced and unvoiced segments considering short term energy and fundamental frequency (F0), ZCC etc. Spectral features (frequency domain features) were also used to classify depressed from control. The spectral features considered in this research are MFCCs and linear predictive group delay. Features normalization was used to reduce the features' range mismatch of different speakers. The classification accuracy with the Gaussian Mixture Model (GMM) was found to be 80% considering MFCC as a feature for the speaker-dependent case and 77 for the speaker-independent case. Alghowinem et al. [22] extracted low-level descriptors features from a frame duration of 25 ms. They have evaluated linear features: fundamental frequency (F0), intensity, loudness, voice probability, quality, jitter, shimmer, HNR, log energy, root mean square energy, and the non-linear teager energy to understand the difference in voiced/unvoiced and mixed speech on depression detection. They suggested the suitability of Teager Energy Operator (TEO) based features for depression detection using voiced and unvoiced speech. Voicing probability and log energy were found to be more suited for mixed speech. They investigated high jitter, lower shimmer, high HNR, lower vocal energy in the glottal pulse of depressed subjects, and lower range of fundamental frequency (F0). The GMM was used for dimension reduction and then Support Vector Machine (SVM) was used to classify depressed and control subjects. Scherer et al. [17] employed a multimodal framework in which video features along with acoustic features were used to classify depressed individuals. The acoustic features that were extracted mainly focused on calculating breathiness and tenseness of voice. NAQ and QOQ, both, are derived from amplitude measurement of glottal flow pulse. It has been deduced that both measures are inversely correlated with the tenseness of the voice. The smaller the value the more tense the voice will be and thereby are highly correlated with each other. Ozdes et al. [34] explored the significance of vocal jitter and spectral slope as indicators of suicidal tendencies. The pairwise classification accuracies among control/depressed, depressed/suicide, and control/suicide using jitter were 0.65, 0.60, and 0.80, respectively, and using spectral slope were 0.90, 0.75, and 0.60, respectively. Using both jitter and spectral slope, the accuracies reported were 0.90, 0.75, and 0.85, respectively, for the three pairs. Sethu et al. [35] evaluated emotions based on pitch, energy slope, and formants for speaker-dependent and speaker-independent studies. MFCCs and LPC-based group delay performed well in the speaker-dependent system, while the first three formants performed well in the speaker-independent system. The study by Scherer et al. [8] involved the analysis of NAQ, QOQ, peak slope, and Open Quotient Neural Network (OQNN). It was observed that these features are independent of gender. 75% accuracy was obtained with SVM as a classifier. In 2016, Pampouchidou et al. [30] used a fusion of low-level features and Discrete Cosine Transform (DCT) based features and high-level features. They analyzed two sets of sizes 494 and 1,278, respectively, using the Distress Analysis Interview Corpus dataset Wizard-of-Oz (DAICWOZ) dataset and the Collaborative Voice Analysis Repository for Speech Technologies (COVAREP) feature repository. The gender-based results using low-level descriptors are reported in terms of f1 depressed (non-depressed) as 0.45 (0.85) (Leave-one-out Cross-validation (LOOCV)) and 0.59 (0.87) (testing using development set). The DCT-based features for gender independent data gave 0.19 (0.71) and 0.47 (0.83) in terms of f1 measure, respectively. In 2017, Pampouchidou et al. [35] evaluated the Continuous Audio/Visual Emotion and Depression Recognition Challenge dataset [26] using the COVAREP-based audio features and reported a precision of 0.948 while using visual OR (ed) with audio gender-based features. Audio alone gave the f1 score of 0.641. In 2017, Cummins et al. [36] performed depression detection using eGeMAPS, COVAREP, gender-dependent VL-formants, and a fusion combination. Performance was measured in terms of f1-score for depressed (non-depressed) classes. The overall f1 depression was calculated to be 0.63 (0.89). Results were performed on the training and development partitions of the DAICWOZ Corpus. All classifications were performed using the Liblinear package [37] using grid search. In 2018, Takaya Taguchi et al. [31] investigated the second dimension of the MFCCs for depression detection. They found it highly discriminatory for classifying depressed subjects from control. An accuracy of 81.9 was obtained using the second dimension of MFCCs feature as a biomarker.

14.8 EXPERIMENTATION ANALYSIS

The quantum computing-based whale optimization algorithm has been found to be outperformed by the Whale optimization algorithm. The experiments conducted with four population sizes varied from 1, 10, 20, and 50. The number of generations has been kept fixed at equal to 100. The runs are kept to be 10; and in each run, 10 fold cross-validation has been performed. The average fitness and accuracy have been found to increase with the increase in population size. To reduce the time complexity of the depression dataset (DAICWOIZ) [38], in which there are multiple features and the number of samples is small, to avoid overfitting, we can try applying hierarchical clustering-based on the Euclidean distance. The reduced features are then input to the quantum-based whale optimization algorithm which would further reduce the feature set. The hierarchical clustering will cluster the features. From each cluster, on the basis of t statistics, each representative is selected to make non-redundant and relevant subset of features. The role of the quantum-based whale optimization algorithm further optimizes the feature set on the basis of using fitness value and accuracy. The datasets used in an experiment and their details are presented in Tables 14.1–14.3:

14.9 CONCLUSION AND FUTURE DIRECTIONS

We can first do hierarchical clustering on the features of the depression dataset. Then on those pre-processed feature set, quantum computing whale optimization is applied

TABLE 14.1
Datasets Used and Their Details [1]

Dimension	Domain	Dataset	Class	Samples	Original Features	Preprocessed Features (m)	Features After Clustering
High	Face Image	AR1OP	10	130	2,400	2,400	100
High	Face Image	PIE10P	10	210	2,400	2,400	100
High	Face Image	PIX10P	10	100	10,000	2,400	100

TABLE 14.2
Comparative Performance for High-dimensional Datasets [1]

Domain	Dataset	Average Fitness		Average Accuracy		Average Features	
		WOA	QWOA	WOA	QWOA	WOA	QWOA
1 Image	AR10P	0.0112	**0.007**	99.46	**99.46**	59	**58.3**
2 Image	PIE10P	0.0041	**0.003**	100	**100**	40.6	**36.2**
3 Image	PIX10P	0.028	**0.0043**	99.9	99.7	18.3	**12.8**

The bold values denote the algorithm that has outperformed in terms of fitness, accuracy, and features.

TABLE 14.3

Comparison of the Proposed QWOA with Previous State of the Art/Previous Literature [1]

		QWOA	Literature	QWOA	Literature
Dimension	Dataset	Accuracy	Feature	Percentage	
High	AR10P	**99.84** 98.00	**0.024**	0.063	[39]
High	PIE10P	**100.0** 98.67	**0.015**	0.0167	[40]
High	PIX10P	**99.70** 99.00	**0.001**	0.005	[41]

The bold values denote the algorithm that has outperformed in terms of fitness, accuracy, and features.

to select relevant feature set which would identify depressed people from the control one. As depression is a multidimensional dataset, relevant subset of features is to be picked out from an exhaustive subset of the feature set. The previously explained evolutionary algorithms can be amalgamated with the quantum approach and can be used for feature selection in obtaining an optimal feature subset. Similarly, GA, ACO, and PSOA can be amalgamated with quantum computing and can be used for feature selection in the multi-dimensional dataset.

ACKNOWLEDGEMENT

I, the main author of the paper, would like to thank the authors of the paper "Quantum-based Whale Optimization Algorithm for Feature Selection" – Prof. R.K. Agrawal and Dr. Baljeet Kaur – who involved me in the experimental section of the paper, which helped me gain in-depth knowledge about quantum computing. They have enabled me to write a chapter on quantum computing. I would like to thank Devesh Maheswari and Shikha Sharma who helped me in proofreading the paper. I would also like to thank Dr Adarsh Kumar who gave me the opportunity to write this chapter.

REFERENCES

1. Agrawal, R. K., Baljeet Kaur, and Surbhi Sharma. "Quantum based whale optimization algorithm for wrapper feature selection." *Applied Soft Computing* 89 (2020): 106092.
2. Mafarja, Majdi, and Seyedali Mirjalili. "Whale optimization approaches for wrapper feature selection." *Applied Soft Computing* 62 (2018): 441–453.
3. Zhang, Yudong, Shuihua Wang, and Genlin Ji. "A comprehensive survey on particle swarm optimization algorithm and its applications." *Mathematical Problems in Engineering* 2015 (2015): 931256.
4. Mitchell, Melanie. *An Introduction to Genetic Algorithms.* MIT Press, Michigan (1998).
5. Sun, Jun, Bin Feng, and Wenbo Xu. "Particle swarm optimization with particles having quantum behavior." In: *Proceedings of the 2004 Congress on Evolutionary Computation (IEEE Cat. No. 04TH8753).* Vol. 1. IEEE, 2004.

6. Han, Kuk-Hyun, and Jong-Hwan Kim. "Quantum-inspired evolutionary algorithm for a class of combinatorial optimization." *IEEE Transactions on Evolutionary Computation* 6.6 (2002): 580–593.

7. Dorigo, Marco, and Thomas Stützle. "Ant colony optimization: overview and recent advances." In: *Handbook of Metaheuristics*. IRIDIA, Bruxelles (2019): 311–351.

8. Wang, Guangtao, et al. "Selecting feature subset for high dimensional data via the propositional FOIL rules." *Pattern Recognition* 46.1 (2013): 199–214.

9. Bi, Ning, et al. "High-dimensional supervised feature selection via optimized kernel mutual information." *Expert Systems with Applications* 108 (2018): 81–95.

10. Song, Qinbao, Jingjie Ni, and Guangtao Wang. "A fast clustering-based feature subset selection algorithm for high-dimensional data." *IEEE Transactions on Knowledge and Data Engineering* 25.1 (2011): 1–14.

11. World Health Organization. *Preventing Suicide: A Global Imperative.* World Health Organization (2014).

12. http://www.assocham.org/newsdetail.php?id=4918.

13. http://indiatoday.intoday.in/story/delhi-police-suicide-high-stress-depressionfinancial-issues/1/1069444.html.

14. Spitzer, Robert L. "Values and assumptions in the development of DSM-III and DSM-III-R: An insider's perspective and a belated response to Sadler, Hulgus, and Agich's. 'On values in recent American psychiatric classification'." *The Journal of Nervous and Mental Disease* 189.6 (2001): 351–359.

15. Leentjens, Albert FG, et al. "The validity of the Hamilton and Montgomery-Åsberg depression rating scales as screening and diagnostic tools for depression in Parkinson's disease." *International Journal of Geriatric Psychiatry* 15.7 (2000): 644–649.

16. Beck, Aaron T., et al. "Comparison of Beck Depression Inventories-IA and-II in psychiatric outpatients." *Journal of Personality Assessment* 67.3 (1996): 588–597.

17. Kroenke, Kurt, et al. "The PHQ-8 as a measure of current depression in the general population." *Journal of Affective Disorders* 114.1–3 (2009): 163–173.

18. Cummins, Nicholas, et al. "An investigation of depressed speech detection: Features and normalization." In: *Twelfth Annual Conference of the International Speech Communication Association*. 2011.

19. Scherer, Stefan, et al. "Automatic audiovisual behavior descriptors for psychological disorder analysis." *Image and Vision Computing* 32.10 (2014): 648–658.

20. Lemke, Matthias R., et al. "Spatiotemporal gait patterns during over ground locomotion in major depression compared with healthy controls." *Journal of Psychiatric Research* 34.4–5 (2000): 277–283.

21. Pampouchidou, Anastasia, et al. "Depression assessment by fusing high and low level features from audio, video, and text." In: *Proceedings of the 6th International Workshop on Audio/Visual Emotion Challenge*. 2016.

22. Calvo, Rafael A., et al. "Natural language processing in mental health applications using non-clinical texts." *Natural Language Engineering* 23.5 (2017): 649–685.

23. Scherer, Klaus R., "Vocal affect expression: a review and a model for future research." *Psychological Bulletin* 99.2 (1986): 143.

24. Alghowinem, Sharifa, et al. "Characterising depressed speech for classification." 2013.

25. Meng, Hongying, et al. "Depression recognition based on dynamic facial and vocal expression features using partial least square regression." In: *Proceedings of the 3rd ACM International Workshop on Audio/Visual Emotion Challenge*. 2013.

26. France, Daniel Joseph, et al. "Acoustical properties of speech as indicators of depression and suicidal risk." *IEEE Transactions on Biomedical Engineering* 47.7 (2000): 829–837.

27. Sethu, Vidhyasaharan, Eliathamby Ambikairajah, and Julien Epps. "Speaker dependency of spectral features and speech production cues for automatic emotion classification." In: *2009 IEEE International Conference on Acoustics, Speech and Signal Processing*. IEEE, 2009.

28. Davis, Steven, and Paul Mermelstein. "Comparison of parametric representations for monosyllabic word recognition in continuously spoken sentences." *IEEE Transactions on Acoustics, Speech, and Signal Processing* 28.4 (1980): 357–366.

29. Vidhyasaharan Sethu, Eliathamby Ambikairajah, and Julien Epps. "Empirical mode decomposition based weighted frequency feature for speech-based emotion classification." In: *2008 IEEE International Conference on Acoustics, Speech and Signal Processing*, pp. 5017–5020. IEEE, 2008.

30. Farrús, Mireia, Javier Hernando, and Pascual Ejarque. "Jitter and shimmer measurements for speaker recognition." In: *8th Annual Conference of the International Speech Communication Association; 2007 Aug. 27–31; Antwerp (Belgium)*. pp. 778–81. International Speech Communication Association (ISCA), 2007.

31. Low, Lu-Shih Alex, et al. "Influence of acoustic low-level descriptors in the detection of clinical depression in adolescents." In: *2010 IEEE International Conference on Acoustics, Speech and Signal Processing*. IEEE, 2010.

32. Alghowinem, Sharifa, et al. "Detecting depression: a comparison between spontaneous and read speech." In: *2013 IEEE International Conference on Acoustics, Speech and Signal Processing*. IEEE, 2013.

33. Kane, John, and Christer Gobl. "Wavelet maxima dispersion for breathy to tense voice discrimination." *IEEE Transactions on Audio, Speech, and Language Processing* 21.6 (2013): 1170–1179.

34. Kane, John, and Christer Gobl. "Identifying regions of non-modal phonation using features of the wavelet transform." In: Twelfth Annual Conference of the International Speech Communication Association. 2011.

35. Ozdas, Asli, et al. "Investigation of vocal jitter and glottal flow spectrum as possible cues for depression and near-term suicidal risk." *IEEE Transactions on Biomedical Engineering* 51.9 (2004): 1530–1540.

36. Scherer, Stefan, et al. "Investigating voice quality as a speaker-independent indicator of depression and PTSD." In: Edited by Stefan Scherer, Giota Stratou, Jonathan Gratch, Louis-Philippe Morency, *Interspeech*. University of Southern California, Institute for Creative Technologies, Los Angeles, 2013.

37. Pampouchidou, Anastasia, et al. "Facial geometry and speech analysis for depression detection." In: *2017 39th Annual International Conference of the IEEE Engineering in Medicine and Biology Society (EMBC)*. IEEE, 2017.

38. Fan, Rong-En, et al. "LIBLINEAR: A library for large linear classification." *The Journal of Machine Learning Research* 9 (2008): 1871–1874.

39. Taguchi, Takaya, et al. "Major depressive disorder discrimination using vocal acoustic features." *Journal of Affective Disorders* 225 (2018): 214–220.

40. Cummins, Nicholas, et al. "Enhancing speech-based depression detection through gender dependent vowel-level formant features." In: *Conference on Artificial Intelligence in Medicine in Europe*. Springer, Cham, 2017.

41. Gratch, Jonathan, et al. The distress analysis interview corpus of human and computer interviews. In: *Proc. of LREC*, pp. 3123–3128, ELRA, Reykjavik, Iceland, 2014.

Index

Milton Keynes UK
Ingram Content Group UK Ltd.
UKHW031131141024
449569UK00006B/283